"The lie that faith  than when the affa brilliant Dr. Micha.......... is on the case. If we'd had him as our physics teacher, most of us would be physicists today! Q.E.D."

—Eric Metaxas, *New York Times* bestselling author of *Miracles* and host of the nationally syndicated *The Eric Metaxas Show*

"Michael Guillen has always had a unique ability to make fascinating science topics understandable and relevant to people's lives. *Amazing Truths* is proof of that."

—Joan Lunden

"As a gifted communicator, Michael Guillen makes science easy to understand. He has done it again with his latest explanations of how the Bible and science complement each other. In a profound and beautiful way he helps us realize that science does not contradict the Bible but in fact supports it. A summer 'must read' for all high school grads going to college."

—Dr. Robert A. Schuller

"Michael Guillen bridges the seeming gap between science and faith better than anyone I know."

—Cal Thomas, Syndicated and *USA Today* columnist/Fox News contributor

"An entire generation is being swept up in a false narrative that the Bible and science are incompatible. Too many today, both young and old, believe that faith in the Bible has been discredited by the new religion of science. In *Amazing Truths*, Dr. Michael Guillen dismantles those myths, presenting hard evidence and sound reason to cause skeptics to reevaluate their own belief system and believers to have renewed confidence in the Bible."

—Curtis V. Hail, President & CEO, e3 Partners / I Am Second

"Here is a truly unusual autobiography: an amazing blend of up-to-date science and the world in which we exist as conscious spiritual beings."

—OWEN GINGERICH, Harvard University professor emeritus and author of *God's Planet*

"Must science and faith be at war? No, says physicist Michael Guillen in this delightfully readable yet thoroughly researched book. Whether you are a believer or a doubter, you'll be astonished by *Amazing Truths*."

—HATTIE KAUFFMAN, Former correspondent of ABC and CBS News and author of *Falling Into Place*

"The title of this book says it all: Michael Guillen illustrates how science and the Bible are partners in revealing amazing truths about life—and life eternal."

—SPENCER CHRISTIAN, Weather forecaster/interviewer/host for KGO-TV—formerly at *Good Morning America*

"For many people today, even self-described Christians, science has replaced the Bible as the ultimate authority in their lives. In *Amazing Truths*, Michael Guillen explains brilliantly why that secular view is unfounded and offers one that is better informed intellectually and spiritually."

—REV. DR. GREG HUGHES, Senior Pastor, Malibu Presbyterian Church

"Michael is one of the most original thinkers I've ever known. He lives where the creative meets the intellect. I can think of no one more qualified to write this book!"

—DAVE ALAN JOHNSON, Writer/producer/director (*Doc, Sue Thomas FBEye, October Baby, Mom's Night Out, Coffee Shop, Woodlawn*)

AMAZING
TRUTHS

LITERARY REPRESENTATION BY WES YODER,
AMBASSADOR LITERARY AGENCY

Also by the Author

Bridges to Infinity: The Human Side of Mathematics

Five Equations That Changed the World: The Power and Poetry of Mathematics

Can a Smart Person Believe in God?

AMAZING
TRUTHS

HOW SCIENCE AND

THE BIBLE AGREE

DR. MICHAEL GUILLEN

ZONDERVAN

Amazing Truths
Copyright © 2015 by Michael A. Guillen, PhD

Requests for information should be addressed to:
Zondervan, 3900 *Sparks Drive SE, Grand Rapids, Michigan* 49546

Zondervan titles may be purchased in bulk for educational, business, fundraising, or sales promotional use. For information, please email SpecialMarkets@Zondervan.com.

Library of Congress Cataloging-in-Publication Data

Guillen, Michael.
 Amazing truths : how science and the Bible agree / Michael Guillen, PhD.
  p. cm.
 Includes bibliographical references.
  ISBN 978-0-310-34375-2 (hardcover) — ISBN 978-0-310-34376-9 (ebook) — ISBN 978-0-310-34545-9 (mobile app) 1. Truth — religious aspects — Christianity. 2. Truth. 3. Philosophical theology. 4. Bible and science. 5. Religion and science. 6. Christianity. Title.
BT50.G845 2016
220.8'5 — dc23   2015032737

All Scripture quotations, unless otherwise indicated, are taken from The Holy Bible, New International Version®, NIV®. Copyright © 1973, 1978, 1984, 2011 by Biblica, Inc.® Used by permission of Zondervan. All rights reserved worldwide. www.Zondervan.com. The "NIV" and "New International Version" are trademarks registered in the United States Patent and Trademark Office by Biblica, Inc.®

Any internet addresses (websites, blogs, etc.) and telephone numbers in this book are offered as a resource. They are not intended in any way to be or imply an endorsement by Zondervan, nor does Zondervan vouch for the content of these sites and numbers for the life of this book.

All rights reserved. No part of this publication may be reproduced, stored in a retrieval system, or transmitted in any form or by any means — electronic, mechanical, photocopy, recording, or any other — except for brief quotations in printed reviews, without the prior permission of the publisher.

Author is represented by Ambassador Literary Agency, Nashville, TN.

Cover design and illustration: Faceout Studio, Tim Green
Interior design: Denise Froehlich

Printed in the United States of America

For my grandparents:
the Reverend Dr. Miguel and Delia Guillén
the Reverend Guadalupe and Michaela Armendariz

And for my mother-in-law:
Barbara Aragon

*Men and women
of enormous faith and intelligence
who have my undying respect
— with love and gratitude*

CONTENTS

Acknowledgments 9

1. Best of Both Worlds: Objective Truth Exists 11
2. Beyond Circular Reasoning: Time Is Linear 24
3. I Am Who I Am: An Entity Can Have Contradictory Natures 38
4. Seeing in the Dark: Significant Parts of Reality Are Hidden from Us 52
5. Not of This World: Light Is Unearthly 70
6. An Egg-straordinary Event: The Universe Was Created Ex Nihilo 84
7. The Certainty of Uncertainty: Truth Is Bigger Than Proof .. 100
8. La Vida Loca: Cause and Effect Can Be Disproportional ... 118
9. The Cosmic Grapevine: Instantaneous Communication Is Possible 135
10. Beyond Fleas and Grapes: Humans Are Unique 154

Notes 173
Scripture Index 185
Name Index 187
Subject Index 190
About the Author 197

ACKNOWLEDGMENTS

ABOVE ALL, I THANK GOD for prompting me to put into words the life-changing insights he has given me during my time as a physicist and, more recently, a Christian. And for the precious gifts, rare opportunities, and invaluable experiences he has granted to me over the years. *Soli Deo gloria* — to him alone be the glory!

I wish to thank my wife, Laurel, who encouraged me to write this book and who read the first draft of each chapter and reviewed carefully the final manuscript, never failing to counsel me wisely on how to make it better. She lovingly put up with my spending countless hours, day and night, holed up in my man cave merrily writing away.

Also, heartfelt thanks to my manager, Burton Taylor, who urged me early on to write this book, even when I wasn't feeling it. His persistence and encouragement ultimately helped win the day.

To Wes Yoder, my literary agent and good friend, I wish to say, "Thank you, brother." Your support of this project not only helped it come about but filled me with optimism about its ultimate success.

Thank you to two distinguished friends for their invaluable comments: Dr. Denis Alexander, a molecular biologist and emeritus director of the Faraday Institute for Science and Religion at St. Edmund's College in Cambridge, England, and the Reverend Dr. Emmanuel Haqq, a theoretical physicist and the senior pastor of Christ Community Church in Belchertown, Massachusetts.

Thank you also to Senior Editor-at-Large Jim Ruark. His vast experience in reading and critiquing manuscripts, as well as his unblinking commitment to excellence, helped make this the best book possible.

Finally, I wish to thank John Sloan, Executive Editorial Specialist, whose great enthusiasm for this book infected me with energy — and whose widely known and respected editorial expertise made me feel confident that the project was in very good hands. Because of his brilliant efforts, you hold in your hands the best possible rendering of the message God intended me to deliver.

CHAPTER 1

BEST OF BOTH WORLDS

OBJECTIVE TRUTH EXISTS

Science without religion is lame;
religion without science is blind.
ALBERT EINSTEIN

Faith and Reason are like two wings
on which the human spirit rises to
the contemplation of truth.
POPE JOHN PAUL II

WHEN I WAS A KID, one of my favorite riddles went like this:

> A boy journeying through a foreign land arrives at a fork in the road guarded by two men in native dress. One man belongs to a tribe that always lies; the other, to a tribe that always tells the truth — but the boy doesn't know which is which. One road leads to danger; the other, to safety. Riddle: What is the one question the boy can ask the two tribesmen to make sure he takes the right road?

I will give you the answer later in this chapter, but first I want to ask you another question. It is the simple but deep question that Pontius Pilate put to Jesus, who claimed to represent the truth: *"What is truth?"*

When I was science editor for *Good Morning America*, I did a story about a tourmaline mine in Southern California. The best part of the assignment was that I got to go mining for gems, something I'd never done.

I was led deep inside the earth to a cramped space where the air was cool and humid. With my bare hands, I dug into the cave's creamy mud wall and soon struck something hard and sharp-edged. I seized the mystery find, pulled it out, and wiped it off. That's when I saw it: a magnificent, fully formed, pink tourmaline gemstone.

I imagine truth — *objective* truth — being like that. Something real and beautiful. Something natural, not of our making, with definable qualities, waiting to be unearthed.

Do such truths actually exist?

I claim they do. And in this book, I elaborate on ten objective truths that I consider amazing because of what they tell us about life, the world, and the Creator. And because — wonder of wonders — science and the Bible *agree* on all ten being absolutely, positively, demonstrably true.

Obviously, my position is at odds with skeptics who believe that science and the Bible cannot possibly agree on anything, and with relativists who insist that objective truth doesn't exist, not in science and certainly not in religion. But the evidence for my thesis is considerable, as you will see.

Why does it matter that objective truth exists? For one thing, because the journey of life is noted for its vexing forks in the road. Whenever we come to one, we are forced to choose which path to take. And our choices are based on ... what? Ideally, I submit, on truth. Objective truth — there is no better GPS than that.

However, if objective truth doesn't exist — if there is nothing and no one totally reliable to help guide our steps — then there is no absolute distinction between the road to safety and the road to danger. If truth is relative, *nothing* has any absolute meaning.

Amazingly, science and the Bible believe in the existence of objective truth. Each believes in the existence of a strict, inflexible criterion — an *axis mundi,* a supreme court, a final, transcendent referee — that judges matters objectively. Each in its own way attempts to elevate us above the din of mere opinion by offering a clear-eyed view of reality — the two distinct views combining to produce a single, majestic, stereoscopic panorama.

For science and the Bible, "right" and "wrong" are not arbitrary. They are based on physical or spiritual laws upheld by formidable evidence and the test of time. Laws that are natural and not man-made. Laws that help us choose the right paths in life. Laws every bit as dazzling as gemstones — and infinitely more precious.

Science

There is a great deal scientists don't understand. The highly acclaimed American physician and author Lewis Thomas put it this way: "The only solid piece of scientific truth about which I feel totally confident is that we are profoundly ignorant about nature."[1]

Moreover, there are times when scientists, being human, are not truthful. In 1974, during Thomas's tenure as president of New York's Memorial Sloan Kettering Cancer Center, an immunologist there named William Summerlin was caught cheating on certain skin transplant experiments. He used a felt-tip pen to make it look as if skin from a black mouse had been successfully grafted onto a white mouse.[2]

Notwithstanding these weaknesses, science represents an extraordinary, centuries-long, collective effort to discover objective truths. At one time or another, individuals from virtually every part of the globe have contributed to its advancement. But one particular people — the Greeks of the sixth, fifth, and fourth centuries BC — is rightfully credited with getting it started.

Historians call it the "Greek Miracle," a series of game-changing insights and discoveries about the physical world made by extraordinary persons such as Thales of Miletus, Socrates, Plato, and Aristotle. The complete list is long and impressive.

The Greek Miracle happened a long time ago, but it came home to me one afternoon when I was in Greece, shooting a segment for

Where Did It Come From?, my History Channel television series. During a break in the filming, I stood on the beach of a remote island and gazed out at the Aegean Sea. Small isles gathered in the near distance were partially shrouded in haze, giving them an otherworldly appearance. I mused about how this remote venue had given birth to revelations — objective scientific truths — that drastically altered the trajectory of human history. I wondered, *Why here? Why the Greeks? Was it something in the water?*

Most notably, the ancient Greeks were the first to come up with the idea of a universe — a *kosmos*, Pythagoras called it — that wasn't nearly as wild and crazy as humans had always thought. It was, instead, a thoroughly rational world that, given time, our brains could figure out.

In order to make sense of nature's dazzling variety, Aristotle lumped together plants and animals that shared certain important physical traits: body type, behavior, habitat, and so forth. (For more on this, see chapter 10.) It is like what many kids do when, after trick-or-treating, they sort through their booty and separate it into piles of candy types. With that simple, systematic technique, Aristotle founded the discipline of biology.

Euclid brought the same semblance of order to the infinite variety of numbers and shapes. He gathered together countless practical mathematical formulas that had been used for centuries by farmers, astronomers, and surveyors and fitted them together into a single, logical body of knowledge. Thus was plane geometry born — to this day, taught in high schools all over the world. One of the objective truths Euclid proved is this: the interior angles of any triangle on a flat plane — be it fat, skinny, long, or short — always add up to 180 degrees. It is not a relative truth, not an opinion shaped by cultural, political, or gender biases. The total of the angles is 180 degrees, period.

Last century, after Albert Einstein articulated the special and general theories of relativity, many people took it to mean that *everything* is relative, including truth. They couldn't be more mistaken. According to Einstein's special theory of relativity, different observers will routinely disagree on superficial realities, such as distances and times. But the laws of physics themselves are not a matter of opinion; they are not relative. They are *identical* for everyone. They are objective truths — Lorentz invariant, to use the lingo of special relativity. An illustration of this fact is simple: leap from atop a skyscraper and you will fall with increasing speed. No matter who you are, no matter what you believe is or is not true, count on it: in the end, you *will go splat*.

Likewise, in life we can disagree on who is the greatest athlete, fashion designer, or pop singer of all time. And we can back up our opinions with evidence. But either the earth circles the sun or it doesn't; either atoms exist or they don't; either the continents drift or they don't. These things are objective truths or they aren't, no matter what anyone says.

Generally speaking, in science, objective truths come in several guises.

First, they can appear in the form of *laws*. In addition to Newton's well-known law of gravity, we have Coulomb's law of electrostatic attraction and repulsion, the law of entropy, the law of energy conservation, the law of action and reaction, among many others.

Second, objective truths can appear in the form of fundamental *physical constants* — the vital statistics of material reality. To name just a handful, we have Newton's constant of gravity (6.673×10^{-11} newton meters-squared per kilogram-squared), the electric constant (8.854×10^{-12} farads per meter), Planck's constant (6.626×10^{-34} joule-seconds), and the speed of light in a vacuum ($299{,}792{,}458$ meters per second). These numerical values and their decisive influence on how things work in this world are not matters of opinion; they are absolutely true for anyone anywhere in the universe.

Third, objective truths can appear in the form of *theories*. Examples are the special theory of relativity, general theory of relativity, quantum theory, big bang theory, inflationary theory, theory of thermodynamics, theory of evolution, and so forth. Theories are far and away the most volatile among the three categories of absolute truth;

they are subject to change, even to being overturned. But so long as they are supported by the latest evidence, they are universal verities — and science acknowledges them as such.

Whatever form they take, objective truths in science deal strictly with the *physical-material* behavior of the universe and everything in it. By its own admission, science is neither qualified nor designed to settle age-old questions about any phenomena that might (or might not) exist beyond the physical-material. I refer here to questions such as these: Does God exist? What's the point of the universe? Of life? Is there an afterlife? A heaven? A hell?

This wasn't always the case, by the way. The earliest scientists — among them devout Christians, Jews, and Muslims — felt quite comfortable including God in their hypotheses, invoking him as the final cause of all physical phenomena. They commonly viewed experiments as rigorous ways to elucidate scriptural truths and were confident that science was affirming God's existence.

Typical of that widespread sentiment were the words of Sir John Herschel, a late eighteenth- and early nineteenth-century English-born astronomer in England: "All human discoveries seem to be made only for the purpose of confirming more strongly the truths come from on high, and contained in the sacred writings."[3]

By Charles Darwin's time, science had largely abandoned its historic role as God's champion and, more than that, had avowed its fundamental incapacity to pass judgment on religious matters, one way or the other.

Today orthodox science outright excludes God from the scientific method — a self-imposed restriction that upsets many people but shouldn't. Like any human enterprise, mainstream science is entitled to define the rules by which it plays. Disagree with them, disobey them, if you wish, but you will not be considered part of the orthodoxy if you do.

To reiterate: modern, mainstream science excludes supernatural explanations not because it has decided that God doesn't exist or proven that physical-material reality is all there is (something that truthfully it can never do); it does so simply because the aim of modern science is to find strictly rational explanations for the natural world.

Not only is that a reasonable thing to want to do, but I submit that it is fundamentally compatible with religion generally and the Bible specifically. In fact, a Christian could argue — as I do — that science's success at explaining the natural world rationally is *guaranteed* because the natural world is the creation of a rational God.

The Bible

Imagine how frightening and unintelligible the untamed world must have seemed to our earliest ancestors. Biblically speaking, this was the period immediately following the fall. A time when, having been banished from the garden and God's proximity, humans found themselves in a foreign, hostile environment.

In the account of Cain and Abel, we catch a flavor of that forbidding, godless world. After slaying his brother, Cain faces exile to a place even farther removed from God and the garden. He cries out to God: "I will be hidden from your presence; I will be a restless wanderer on the earth, and whoever finds me will kill me" (Gen. 4:14).

During the ages that followed, as we, the physical, emotional, and spiritual descendants of Adam and Eve, explored our scary new environs, we felt the urge to ask not only *what and where* but also *who and why* — a prompting, I believe, from some sort of vestigial recollection or genetic-like memory of our godly origins. (For more on this, see chapter 10.)

To a rational-materialist, our metaphysical proclivities indicate there is something very wrong with us — that our stubborn impulse to believe in a spiritual reality is the symptom of a mental weakness, the outward expression of a mass delusion. I myself believe just the opposite — that there is something very wrong with glibly dismissing a trait so clearly fundamental to who we are, a powerful human instinct that perceives the existence of something beyond the realm of our provincial five senses.

In his excellent book *Instinct*, T. D. Jakes likens our situation to a lion at the zoo that somehow knows there is a world beyond the locked cage, a lion that will escape from the cage the first chance it gets. "Our instincts," Jakes says, "are the treasure map for our soul's satisfaction."[4]

My disagreement with rational-materialists aside, the amazing truth is this: from the get-go, we humans have felt compelled to explain the cosmos using not just temporal-physical language but metaphysical language as well. Before the founding of science, before the Greek Miracle, individuals from virtually every part of the world participated in the attempted explanation. Gradually they pieced together metaphysical worldviews — among them, what we now call Hinduism, Buddhism, and Judaism — that attracted first local and eventually large regional followings.

But then something happened in Israel in 2 or 3 BC that changed everything forever. I call it the "Christian Miracle," a series of game-changing disclosures about the metaphysical world by extraordinary persons such as Mary, Joseph, John the Baptist, and, above all, Jesus of Nazareth. I find it an interesting coincidence that the Christian Miracle happened not far from the Greek Miracle. Bethlehem is located just across the Mediterranean Sea from Athens, a mere 785 miles away. And like the Greek Miracle, the Christian Miracle comprises revelations given to a certain people that eventually spread throughout creation. Without the help of Twitter or Facebook or Instagram, Christianity went viral in the span of several centuries.

The Christian Miracle's chief disclosure is this: the Creator of the universe is not at war with us. Neither is his favor for sale; no number of brownie points can possibly buy it. God loves us freely and unconditionally.

According to Christianity, that good news is not merely an opinion. It is an objective truth applicable to everyone everywhere for all time:

> "For God so loved the world that he gave his one and only Son, that whoever believes in him shall not perish but have eternal life." (John 3:16)

I will never forget my first trip to Israel. I checked into the King David Hotel in Jerusalem and stayed in a room facing the walled city. That first night, as I gazed out my window at the gate Jesus reportedly walked through to celebrate his final Passover on earth, I was overcome with a feeling similar to the one I'd had looking out at the Aegean.

This is a chosen place, I thought, *where great objective truths about spiritual reality were revealed and where our final destiny will one day play out.* It is a powerful feeling I experience every time I return to Israel.

Powerful. That is the best way to describe objective truths. Truths aren't objective because someone in authority says they are. They have a power all their own that delivers the goods over the long haul — be it in the controlled experiments of science or the dramatically rehabilitated lives of ordinary people. Objective truths have a unique persuasiveness that enables them to outmuscle trendy thoughts and cherished behaviors — even stubborn ones such as slavery in Africa, Europe, and America. It is how and why the followers of Jesus have endured — are, in fact, flourishing — long after the religions formed by adherents of countless other putative messiahs have died out. It is how and why on February 27, AD 380, Christianity was able to defeat unbelievably violent skepticism and persecution and become the official religion of the Roman Empire.

Powerful. "No army can withstand the strength of an idea whose time has come," Victor Hugo once said. To which I would add the words of the legendary Baptist preacher Charles Spurgeon: "The Word of God is like a lion. You don't have to defend a lion. All you have to do is let the lion loose, and the lion will defend itself."

Christianity is now the world's most practiced religion. According to the Pew Research Center's 2012 report *The Global Religious Landscape*, 2.2 billion people — about 31.5 percent of the world's population — are Christians.[5] Moreover, Christians are so widespread that "no single continent or region can indisputably claim to be the center of global Christianity."[6] That cannot be said of any other religion.

Christianity is also unique among religions in its strong defense of objective truth. By contrast, many Eastern religions tend to believe that truth and reality are illusory. "In a real sense," explains the Tibetan meditation master Namkhai Norbu, "all the visions that we see in our lifetime are like a big dream."[7]

Such mystical thinking favors a solution to my opening riddle that is not in line with Christianity or science. It holds that the road to safety and the road to danger are both just in our heads, that so long as we maintain a correct frame of mind, it doesn't matter which path we trod; therefore we don't need to worry about what question to ask the pair of natives. According to Christianity, however, the two roads represent choices in life that are truly distinct and very real — a view wholly in line with science's belief in objective reality.

So what is the solution to my riddle? Ask the two natives the following question: *"If* I were to ask you which road leads to safety, what would you say to me?" The honest native will give you the correct answer. And so will the liar — because he will be forced to lie about the lie he normally would have told you. A double negative is a positive!

In the riddle of life, too, Christianity believes that it is absolutely, objectively true that there exists a correct path and an incorrect path, and that the solution to that riddle will put you on the correct one. Nowhere else is this uncompromising position on objective truth more dramatically illustrated than in Jesus' table conversation with his disciples during the Last Supper:

> Jesus: "Do not let your hearts be troubled.... You know the way to the place where I am going."
>
> Thomas: "Lord, we don't know where you are going, so how can we know the way?"
>
> Jesus: "I am the way and the truth and the life. No one comes to the Father except through me. If you really know me, you will know my Father as well. From now on, you do know him and have seen him." (John 14:1–7)

Jesus was voicing not merely an opinion here but an objective metaphysical truth — a truth that is, when all is said and done, Christianity's greatest contribution to our understanding of the spiritual world. According to this powerful, objective truth, to know Jesus is to know God himself: the God of life, the God of light, the God of love.

What Does It All Mean to You and Me?

In 2012 American bicyclist Lance Armstrong was outed for cheating in the Tour de France not once but *seven* times. Besides being publically disgraced, his championship titles were revoked and he was banned from the sport for life.[8]

Think about how different things would be if relativists were correct and truth were only what each of us made it out to be. The reaction to Armstrong's behavior might go something like this: "So he broke the rules of the competition. But who's to say he did anything *wrong*? Your idea of truth is always to play by the rules. His truth is to do whatever it takes to win. So stop hating on him, people!"

If that sounds a bit far-fetched, it is not because I mischaracterized the relativist position. I didn't. It is because such a defensive reaction doesn't ring true. Something about it sounds wrong.

As we have seen, the universe gives every indication of revolving not around human opinions but objective truths — which both science and the Bible are in the business of bringing to light. That, I believe, is the first of two important practical lessons we should learn from this discussion.

I say "practical," because this uncompromising devotion to truth is the chief reason we hold people accountable. Being the inhabitants of a truth-centered universe, we tend to come down hard on anyone who appears to violate certain objective standards of proper behavior, on those who lie, cheat, or bend the truth for personal gain. Just ask Lance Armstrong.

The universe itself punishes us if we thumb our noses at objective truth. Toy with nuclear energy in the belief that it is harmless and very quickly we will be set straight.

If we don't respect it, objective truth will hurt us.

However, if we do respect it, objective truth will ultimately vindicate us and set us free — though not necessarily in our lifetime. That is the second lesson for us.

In science, many alleged heretics have not lived long enough to see their views justified by objective truth. Alfred Wegener for one. In the early 1900s, the German geophysicist was derided by colleagues for proposing that continents constantly shift positions. Now it is the mainstream view.

"The battle of the theory of continental drift is one of the classic tales in the history of science," explains Danielle Clode in her excellent book *Continent of Curiosities*. "Wegener did not live to see his work vindicated, dying in the pursuit of science on the ice cap of Greenland."[9]

In religion, Solomon, writing in Ecclesiastes, complains about the same thing. "In this meaningless life of mine I have seen both of these: the righteous perishing in their righteousness, and the wicked living long in their wickedness" (7:15). Elsewhere in the same book, however, Solomon acknowledges that even if justice isn't achieved in this life, it is in the next. Objective truth will out. "God will bring into judgment both the righteous and the wicked, for there will be a time for every activity, a time to judge every deed" (3:17).

How are *you* electing to live? As if objective truth exists or as if all truth is relative?

After Lance Armstrong was found out, Jan Ullrich, his chief rival, felt compelled to confess his own illegal use of the performance-enhancing substance EPO, short for erythropoietin. But did he admit he had done anything wrong? Was he contrite? *No.* Here is what he said: "In my view you can only call it cheating on my part when it is clear that I have gained an unfair advantage. That was not the case. All I wanted was everyone to have the same chances of winning."[10] For Ullrich, in other words, two wrongs make a right. That's his truth.

Does your rhetoric resemble Ullrich's? If so, chances are you are living the life of a relativist even if you don't consciously buy into relativism.

It doesn't have to be that way, you know. As we will see in the forthcoming chapters, life is not about *proving* that objective truth exists — that's not always possible to do, and certainly not by using logic alone. Life is about *believing* that it exists.

Like science and Christianity, I believe that objective truth exists, and I try to live accordingly. I have cheerfully dedicated my life to mining for those amazing insights about creation and the Creator that dazzle the eyes with their beauty and brilliance. That's why I wrote this book. Not to convert you to my way of seeing things — I don't have that kind of power. But to explain the reasons why I rejoice daily in the choice I have made — and what a huge, practical difference that choice has made in my life and can make in your life as well.

CHAPTER 2

BEYOND CIRCULAR REASONING

TIME IS LINEAR

*Jesus Christ is the same yesterday
and today and forever.*

HEBREWS 13:8

Past time is finite; future time is infinite.

EDWIN HUBBLE

TIME IS LIKE THE OPPOSITE SEX. It is at once greatly familiar and profoundly mysterious.

During my graduate studies in physics, math, and astronomy at Cornell, I was clueless about male-female relations. I had no social life and certainly no girlfriend. I was such a geek that my idea of a good time was hanging out in my man cave at the Lab of Nuclear Studies or in the stacks of Olin Library.

Back then, being far more interested in studying time than dating women, I designed and taught an undergraduate course on the subject that became quite popular with the students. In retrospect, I see that my interest in time prepared me perfectly for married life. Whether studying time or engaging in the wonderful and vexing experience of being intimate with another person, you never really, fully understand either one.

I have been married twenty-four years to the same beautiful woman, and yet her day-to-day behavior — her reactions to movies, to shopping, to rearing our teenage son — can still leave me shaking my head. And just to be fair, men's behavior — mine in particular — can be just as puzzling and oftentimes is just as infuriating to her.

"What's up with men?" she'll ask.

"What's up with women?" I'll say.

It is the same frustration we all feel when trying to take hold of that slippery, slippery serpent called time.

"What then is time?" mused the brilliant fifth-century saintly scholar Augustine of Hippo. "If no one asks me, I know what it is. [Yet] if I wish to explain it to him who asks, I do not know."[1]

The question Augustine was contemplating is huge and not at all academic. By asking it, he tackled head-on a mystery that is joined at the hip with the very deepest questions we are wired to ask:

- What is life?
- What is death?
- Is there life after death?
- Why must we die?
- What does it mean to be immortal?
- Where exactly in time does God and heaven and hell exist?

"All people are like grass, and all their glory is like the flowers of the field; the grass withers and the flowers fall, but the word of the Lord endures forever" (1 Peter 1:24–25). In that oft-quoted passage from 1 Peter, we see that the Bible, like science, believes that time is linear, that it resembles a sequenced series of numbers extending from zero to positive infinity — and that we experience it in only one direction. Once we are born, there is no going backward.

The antithesis of this belief is that time is circular, that you and I wheel round and round in a temporal circle, or spiral, until or unless we somehow manage to escape from it. It is like what Bill Murray's weatherman character experienced in the movie *Groundhog Day*. He was doomed to live and relive the same day until his curmudgeonly behavior shaped up.

The circular view — by far the oldest of the two competing models of time — is espoused by Buddhism and other religions we commonly associate with the East. Gautama Buddha, born around 560 BC in northeastern India, now Nepal, is reported to have spiraled round and round the great Circle of Life many times before finally getting it right. To be precise, he made *five hundred* full circuits of terrestrial toil and trouble before at last he achieved enlightenment and nirvana.

The disciples of circular reasoning believe that we are given the opportunity to reenter the highway of life over and over again — each time representing a new opportunity to navigate by our own lights to some spiritual paradise. It sounds appealing — who doesn't relish getting a second chance at a nice reward? But the penalty for getting it wrong anywhere along the way can be severe.

In modern-day Tibet, for example, blind and otherwise congenitally disabled children are treated cruelly by society, even spit at and cursed publically. In the documentary *Blindsight*, a woman is heard shouting at two blind kids who accidentally bump into her on the street: "Look out, morons. You deserve to eat your father's corpse!"[2]

What's behind this unspeakable behavior? Buddhism's belief in circular time and *karma*, in particular the belief that blind children are reincarnations of people who messed up in a previous life. In the aforementioned documentary, one of the children ponders aloud: "It's because of my bad deeds in a previous life that I'm blind in this one. It's what's written in my karma."[3]

Science and the Bible offer us a worldview very unlike the East's hamster wheel of life. According to science and the Bible, time, and therefore life, advances in a straight line from the past, through the present, and onward to the future. The journey can be dangerous and terrifying. But to science and the Bible, as we are about to see, linear time is, above all, the stuff of endless possibilities and soaring creativity.

The Bible

Christianity pioneered the Western view of linear time. "It is to Christianity that we owe our modern temporal orientation," writes British mathematician and science historian G. J. Whitrow in *Time in History*.[4] He quotes German philosopher Erich Frank as concurring: "With Christianity ... man acquired a new understanding of time."[5]

Long before anyone else, including scientists, Christians introduced the world to the idea that time advances in a straight, orderly fashion toward a knowable end. That time is the thread of a universal story line — to many, the greatest story ever told.

The radical nature of Christianity's temporal worldview becomes obvious when we look at the history of religious thought, starting with the development of Hinduism in India. Arguably the world's oldest major religion, Hinduism in recognizable oral form could well date to 4000 BC. Back then, like virtually all of their contemporaries, Indians believed in circular time. What's more, because they didn't see the world in terms of a rectilinear past, present, and future, they usually weren't motivated to record dates. As the German idealist Georg Hegel once complained: "Nothing can be more confusing, nothing more imperfect than the chronology of the Indians; no people which attained to culture in astronomy, mathematics, etc., is as incapable for history; in it they have neither stability nor coherence."[6]

For this reason, the exact origins of Hinduism itself are shrouded in mystery. From what scholars have been able to piece together (albeit contentiously), it appears accurate to say that Hinduism is the amalgamation of many circular-minded, polytheistic religious traditions associated with the early inhabitants of the Indus Valley. These include Vedism, a belief system possibly belonging to the indigenous people of the valley or to the Aryans, a Caucasian people who might have come from the Russian steppes. As I said, it's all a bit hazy.

The most conspicuous indicator of Hinduism's circular way of thinking is the *samsara* — a belief in reincarnation, in the great Circle of Life. According to this conviction, a newborn's physical being is determined by the number of merits — the quantity of *karma* — he or she has earned during all of his or her past circumnavigations of earthly existence. The more *karma* the child has racked up by thinking, believing, and acting "rightly" — as judged by certain prescriptive behaviors known as *dharmas* — the higher up on the food chain the child's soul or *atman* can expect to land in the next go-round of life.

The renowned Sufi poet Jalâl ad-Dîn Rûmî, though neither Indian nor a Hindu, accurately summarized (albeit with a decidedly positive spin) reincarnation's doctrine of the transmigration of souls: "I died as a mineral and became a plant, I died as a plant and rose to animal, I died as animal and I was man. Why should I fear? When was I less by dying?"[7]

When Hinduism was some four thousand years old, the Christian Miracle happened. That is when our general perception of time really changed, when it went linear *and* universal. Unlike any other religion before it, including Judaism, Christianity is the story of all peoples for all time, past, present, and future. It braids our individual story lines into a single, contemporary, cosmic narrative. It says that from start to finish in this time-driven journey called life, we are all moving forward together in the same, straight direction — not spiraling around in separate, meandering whirlpools.

The great importance that Christianity assigns to linear time is readily seen in the New Testament's meticulous attention to temporal details. One instance of this is in Luke 3:23, where we read that "Jesus himself was about thirty years old when he began his ministry." We see it as well in the specificity of the New Testament's historical references. The census decree by Caesar Augustus that caused Joseph and Mary to travel to Bethlehem. The marriage of Herod Antipas and Herodias, whose daughter Salome demanded the head of John the Baptist. The early first-century tenure of Pontius Pilate, Roman prefect of Judea, who sought to wash his hands of the kangaroo-court proceedings that led to Jesus' crucifixion.

Like any well-written, time-ordered narrative, the greatest story ever told can be divided into three acts, three distinct ages heralded by a prelude.

The prelude comprises many prophecies memorialized in the Old Testament that are destined to be fulfilled by Jesus the Christ. In Micah 5:2, the eighth-century BC prophet recorded this prediction about the Messiah's birthplace: "But you, Bethlehem Ephrathah, though you are small among the clans of Judah, out of you will come ... one who will be ruler over Israel, whose origins are from of old, from ancient times."

Act 1 starts with events leading up to Jesus' birth and ends with his world-changing ministry and sacrifice. "The crucifixion was considered by [Christians] to be a unique event not subject to repetition," explains mathematician and science historian Whitrow. "Consequently, [according to them] time must be linear rather than cyclic."[8] This belief is underscored in 1 Peter 3:18: "Christ also suffered once for sins, the righteous for the unrighteous, to bring you to God. He was put to death in the body but made alive in the Spirit."

Act 2 encompasses the nail-biting times we are experiencing now — when the Enemy is having a field day — and ends with the rapture, tribulation, and second coming of Christ. The sequence of events, the descriptions of what is to come, and the details of the denouement vary according to differing interpretations of biblical eschatology. But here again, however it is interpreted, the New Testament emphasizes linear temporal detail.

The apostle John, in Revelation 11:1–2, predicts that the court of the "temple of God" in the "holy city" will be occupied by foreign powers for exactly "forty-two months." At that time, two prophets will appear, and for "1,260 days" (v. 3) they will warn everyone that the end of the fallen world is nigh. The beast (Antichrist) will then rise up and assassinate the prophets, whose bodies will lie in the streets for "three and a half days" (vv. 7–9) — after which they will be resurrected, a miracle that signals the beginning of the end of the world as we know it (vv. 11–12).

Act 3, as I see it, starts with the final judgment and ends with the fulfillment of the glorious prediction made by John in Revelation 21:1–4: "Then I saw 'a new heaven and a new earth,' for the first heaven and the first earth had passed away.... And I heard a loud voice from the throne saying, 'Look! God's dwelling place is now among the people, and he will dwell with them.... "He will wipe every tear from their eyes. There will be no more death" or mourning or crying or pain, for the old order of things has passed away.'"

So, then, when exactly will God step in, clean up our mess, and deliver on his promise to "make all things new"? Clearly, it is the most important time-related question we can ask as we experience the throes of act 2.

And once again, the answer is found in the time-minded New Testament — although it isn't what most of us want to hear: "About that day or hour no one knows, not even the angels in heaven, nor the Son, but only the Father" (Matt. 24:36).

Only the Father, not any of us. The Father, who created linear time. The Father, who exists outside of time. The Father, who is able to take in, all at once, the endless streaming that we time-bound creatures call the past, present, and future.

Science

Like the earliest religions, science started out with a circular view of time. Nowhere is this better illustrated than in ancient Egypt, starting some five thousand years ago. There we see the first inklings of astronomy, medicine, and civil engineering, as well as the natural origins of circular time-keeping.

The ancient Egyptians lived in a desert climate, so each day they awoke to roughly the same weather forecast: hot and dry. (It reminds me of where I was born, Southern California!) As you can imagine, this unrelenting meteorological sameness did nothing to suggest the existence of a distinguishable past, present, and future. What stood out for the Egyptians — what led them from a static to a circular view of time — were certain conspicuous cyclical phenomena. The first was terrestrial — the seasonal ebbing and flowing of the Nile River, which inspired a civic calendar. The second was extraterrestrial — the nocturnal risings and settings of the heavenly bodies across Egypt's cloudless desert sky, which became the basis for an astronomical calendar.

Because of their cyclical view of time, ancient Egyptians didn't see their calendars as marking time cumulatively, the way we do today. For them, yearly calendars did not build on one another to create a seamless timeline; they represented quite simply a succession of independent cycles.

For this reason, the Egyptians saw the tenures of their kings and pharaohs as repetitive, not as historical periods within one continuous timeline. They reset their calendars to zero every time a new pharaoh was inaugurated. Like the lyrics of the Who's classic song "Won't Get Fooled Again," the tune that ancient Egyptians sang was "Meet the new pharaoh. Same as the old pharaoh."

In classical Greece, too, cyclical time ruled the day and night. Take, for instance, their Great Year. Definitions of it varied from one generation and school of thought to another, but all were based on circular thinking. For Plato, the Great Year represented the time it took for the sun, moon, planets, and stars to cycle away from their original relative positions and back again — a total of about fifteen thousand years. Others read into the Great Year an even deeper significance. For the philosopher Heraclitus, for instance, it represented the period of time when an "old" universe was destroyed by fire and a "new" one sprang into being, like some stupendous phoenix.

Subsequently, Stoic philosophers took the concept to a Hindu-like extreme. They believed that the Great Year was a time when everyone and everything were renewed. As summarized many centuries later by the Christian philosopher Nemesius of Emesa, Stoics believed that "every city and village and field will be restored, just as it was. And this restoration of the universe takes place not once, but over and over again — indeed to all eternity without end. [For that reason] there will never be any new thing other than that which has been before, but everything is repeated down to the minutest detail."[9]

Even Aristotle — the great-grandfather of science — had a soft spot in his thinking for circular time. This proclivity could express itself in startling ways. In *Meteorologica*, for example, he claimed that throughout history the entire corpus of human knowledge — the arts and sciences — had been accumulated and lost many times over. He put it this way: "We must say that the same opinions have arisen among men in cycles, not once, twice, nor a few times, but infinitely often."[10]

It isn't easy to pinpoint when exactly the tide of scientific opinion started to change in favor of linear time. But surely Aurelius Augustinus — the famous hedonist turned Christian who ultimately was canonized — ranks as one of the earliest scholars to articulate what ultimately became the orthodox scientific view of time.

We know a good deal about Aurelius. He was born in North Africa in AD 354 to parents who were well respected but not overly well endowed financially. His father was a pagan, his mother a devout Christian. Both parents had high hopes for their son, who very quickly revealed himself to be intellectually precocious.

Aurelius's brilliance presented the family with a serious challenge because they lived in Tagaste (now Souk-Ahras, Algeria), back then the "boonies" of the sprawling Roman Empire. To provide him with a good education, Aurelius's parents paid to send him to the nearest university town, which kept them penniless or close to it.

Despite his mother's Christian influence, young Aurelius renounced orthodox Christianity and its moral traditions. Eventually relocating to Carthage, he succumbed to big-city temptations and completely gave himself over to heresy and hedonism. When he was eighteen, Aurelius took up with a girl, whose name we do not know, and had a bastard son by her. In that same year, his father died.

In AD 386, while living in Milan, Aurelius had an ecstatic vision that resulted in his abrupt, Pauline-like conversion to Christianity. He resigned his teaching position and eventually moved back to his hometown, firmly committed to serving the Christian God for the rest of his life. A decade later he was appointed bishop of Hippo Regius, a major city in Roman Africa.

During his post-conversion years — which overlapped with Christianity's ascendancy within the Roman Empire and ultimate status as its official religion — Augustine of Hippo, as he came to be called, became a towering figure in the early effort to recast classical Greek philosophy in light of mainstream Christian thinking. This included Christianity's revolutionary linear view of time.

Augustine roundly ridiculed the antiquated notion of circular time. Here is how he put it in his famous opus *The City of God*: "The pagan philosophers have introduced cycles of time.... From this mockery they are unable to set free the immortal soul, even after it has attained wisdom, and believe it to be proceeding unceasingly to false blessedness and returning unceasingly to true misery.... It is only through the sound doctrine of a rectilinear course that we can escape from I know not what false cycles discovered by false and deceitful sages."[11]

Augustine's scholarship helped to elaborate notions about linear time that are now at the heart of modern science. Let me give you three quick examples of what I mean.

First is the idea that time has an existence independent of any specific, natural system of time-keeping. It is a far cry from the ancient Egyptian belief that time had no meaning apart from cyclic phenomena, such as the ebbing and flowing of the Nile or rising and setting of the stars. Augustine postulated that our concept of time was rooted in the mind's native desire to order events. Thus he pioneered ideas that foreshadowed Einstein's special theory of relativity.

Second is the idea that time has an origin; that it hasn't always existed. This is now a central feature of science's big bang hypothesis and its numberless variations.

Third is the idea that time progresses not episodically in fits and starts, but smoothly and continuously. Roughly one thousand years following Augustine, the English philosopher Francis Bacon — one of the founders of modern science—reiterated this same concept, likening time to a flowing waterway. "Time," he concluded, "seemeth to be of the nature of a river or stream."[12]

The concept of linear time is now fully identified with science as we know it. It is built into the mathematical equations and theories that scientists use to articulate their worldview. And linear time is foundational to the experimental process used for discovering objective truth, a process Bacon and others helped to formulate. It was first called the "experimental method," then the "method of science," and finally the "scientific method."

And that isn't even the most amazing part. Not only did science derive its belief in linear time from the Bible, like Christianity, but it also believes that everything in the universe participates interactively in a universal story line — that everything in the universe is heading linearly toward a knowable end.

What end does science foresee for our world? It isn't the vision described in the New Testament exactly, but it is apocalyptic.

According to science and its latest cosmological observations — such as those from NASA's Wilkinson Microwave Anisotropy Probe — the warp and woof of the universe will continue being ripped apart by some unseen force called "dark energy" until finally the cosmos grows completely cold and dark. At that moment, quite literally, it will be "lights out" for the world as we know it — possibly even for time itself.

What Does It All Mean to You and Me?

Science and the Bible agree that we get a single chance at life. There are no mulligans, no possibility of cycling back and trying it again. That is the first lesson for us in this discussion.

Scientists and science-fiction writers are, however, fond of toying with hypothetical paradigms involving time travel. They also like to speculate about multiple universes, multiple variations of you and me having the chance to make different choices in different, parallel worlds. One of my favorite sci-fi stories is *The Time Machine* by H. G.

Wells, in which the Time Traveler travels into the future and even has the ability to influence it. That is way cool.

Another favorite is *Einstein's Dreams*, a fascinating novel by my friend and fellow physicist Alan Lightman. In his story, Alan imagines the ways people's behavior is greatly shaped by the temporal dimensions of their different worlds. There is a world where people live discoordinated lives because time is relative; another where people fear change because time moves at a snail's pace; and another where people languish because they have an unlimited time to live.

Particularly poignant, I think, is the lot of the people who inhabit a world where time is circular — and they *know* it: "In the dead of night these cursed citizens wrestle with their bedsheets, unable to rest, stricken with the knowledge that they cannot change a single action, a single gesture. Their mistakes will be repeated precisely in this life as in the life before. And it is these double unfortunates who give the only sign that time is a circle. For in each town, late at night, the vacant streets and balconies fill up with their moans."[13]

In our world, we get exactly one bite at the apple — as did our physical, emotional, intellectual, and spiritual antecedents Adam and Eve. That's scary. But it is also exhilarating and always novel, rarely predictable and never recycled. In *The Belief in Progress*, Scottish theologian John Baillie expressed it this way: "Through the creative power of God the course of events [i.e., linear time] is characterized by the emergence of genuine novelty."[14]

The scariness and exhilaration of linear time comes home to me every time I do a live shot on television. Like the time I had a chance to make history at ABC News by being the first reporter to broadcast live from Antarctica to the United States. The feat had never been done before because no civilian satellite existed with a southerly enough orbit to transmit a signal from so far down under. But I contacted my buddies at NASA, and they allowed me to use one of their Tracking and Data Relay Satellites (TDRS).

When the big day arrived, the executive producer of *Nightline* back in Washington, DC, instructed my producer and me to pretape a run-through of the show as insurance against something going haywire with the planned live shot. And we did.

Finally, the time came for us to do the real thing. As I stood on the ice facing the TV camera — surrounded by my crew, the scientists I was to interview, and some onlookers from nearby McMurdo Station — I was a jumble of emotions. I felt nervous, excited, awed, and proud of what we were about to pull off.

"Ten minutes!" my producer, Rick Wilkinson III, called out.

I took a deep breath and mentally rehearsed what I planned to say.

"We've got a problem!" Rick shouted suddenly.

I told myself not to be distracted by the commotion that erupted behind the scenes. *Nightline* was an important show, and I didn't want to blow it.

As the clock ticked and the commotion continued, a decision needed to be made: Should we go with the pretaped show? Or did we dare go for the live show in the hope that the problem — whatever it was — would be fixed in time? One way or the other, we needed to be on the air at 11:30 p.m. eastern standard time sharp.

With less than a minute left, Rick announced, "We're a go! Stand by everyone." That evening we made history; the live shot went off without a hitch.

Afterward I asked Rick what the last-minute problem had been. He told me a mother seal had leaped out of a breathing hole in the ice and plopped herself right in the line of sight of a laser beam that was part of the complex communication chain that carried our live signal from the ground up to the TDRS. At the very last minute, our engineering crew had managed to lure her away with fish. Now, that's excitement!

There is a second practical lesson for us here. In the world we inhabit, where time is linear, you and I are uniquely lovely. We are never seen as being a worm or a spider or anything less than human because — to quote Psalm 139:14 — you and I are "fearfully and wonderfully made" by a Creator whose "works are wonderful." I like how this is said in Jonny Diaz's hit song *More Beautiful You*. Its message, though directed at young women, is applicable to everyone: don't change your appearance in order to conform to the glamorous images of those whom the secular world considers beautiful. Each of us is a different likeness of God, so, as Jonny sings it, "There could never be a more beautiful you, more beautiful you."[15]

The third lesson for us is this: heaven is not a gated community. It is open to everyone. Buddhism and Hinduism see time as a spiral staircase that leads to heaven. Only those good enough — who work slavishly over the course of many lifetimes — manage to make the climb, which means that at any given moment, some people are considered better than others. But in this world of linear time, life is a one-shot experience. And heaven is a choice, not a chore. It is a gift from God, not a feather in anyone's cap. Ephesians 2:8–9 explains it well: "For it is by grace you have been saved, through faith — and this is not from yourselves, it is the gift of God — not by works, so that no one can boast."

This amazing truth has huge implications for the kind of society we create and champion on earth. That is the final practical lesson of this discussion. As we have seen, science and the Bible — specifically, Christianity — freed us from circular-minded belief systems and rigid caste systems. To this day, they both inspire us to live up to the words of the Declaration of Independence: "We hold these truths to be self-evident, that all men are created equal, that they are endowed by their Creator with certain unalienable Rights."

The capital letters in "Creator" and "Rights" are critical because they signify that in a world truly consistent with science and the Bible, where time is linear, you and I are the product, not of our *karma*, but of the Creator of the universe. They remind us each day that it is God and his love for us — not our good deeds — that define our fundamental worth as we journey together down the wondrous, ineffable, unpredictable river of time.

CHAPTER 3

I AM WHO I AM

AN ENTITY CAN HAVE CONTRADICTORY NATURES

A complete elucidation of one and the same object may require diverse points of view which defy a unique description.

NIELS BOHR

Christ Jesus: Who, being in very nature God ... made himself nothing by taking the very nature of a servant, being made in human likeness.

PHILIPPIANS 2:5 – 7

"HI, MY NAME IS MICHAEL GUILLEN."

Beneath that familiar salutation is a seemingly trivial assumption, one that is in fact freighted with fascinating, far-reaching complexities. The assumption is that I am — that each of us is — a single, coherent, undifferentiated individual, which is not true. Each of us has more than one identity.

In my case, I'm Dr. Michael Guillen, the award-winning communicator who has written bestselling books, appeared on national television, and taught at one of the world's greatest universities. But I'm also Michael Guillen, the kid born in East Los Angeles who still loves going back to the old 'hood for some *pan dulce* and a good, home-cooked Mexican meal. And I'm Mikey Guillen, the member of a loud, large, and loving extended family. I could go on and on.

In extreme cases, the multiple identities are symptomatic of an enigmatic condition psychiatry labels dissociative identity disorder, or DID. Think Robert Louis Stevenson's *Dr. Jekyll and Mr. Hyde*: "Hi, my name is ... take your pick."

According to the National Alliance on Mental Illness, DID is "a disturbance of identity in which two or more separate and distinct personality states (or identities) control an individual's behavior at different times." The alliance estimates that up to one percent of the general population is afflicted with DID and that each victim, on average, has ten distinct identities or "alters."[1] Oftentimes the identities are remarkably different, even antithetical. For example, docile and violent, educated and ignorant, adolescent and aged.

In *The Minds of Billy Milligan*, bestselling author Daniel Keyes describes a young man born in Florida who became the first person in American history to be acquitted of a violent crime because of DID. Billy Milligan reportedly comprised twenty-four disparate alters, each of whom answered to a different name. Among them were Arthur, an urbane, medically savvy Englishman; Regan, a twenty-three-year-old Yugoslavian who spoke broken English and fluent Serbo-Croat; and Adalana, a teenage lesbian.

DID is clearly a tragic disorder. But both science and the Bible agree on the existence of a very different brand of dissociative identity — one that is far from tragic and far from dysfunctional. One that represents a remarkable, other-worldly truth wherein a single entity appears to embody two completely contradictory yet thoroughly integrated traits.

As we are about to see, this amazing truth has revolutionized our understanding not just of the universe, but of our personal relationship with it, both physically and spiritually.

Science

Our earliest rational-scientific descriptions of the universe attached great importance to the existence of opposites. The pre-Socratic Pythagoreans, for example, saw the world in terms of ten polar opposites — ten being the perfect number, according to the Pythagorean worldview:

Limited	Unlimited
Odd	Even
Unity	Plurality
Right	Left
Male	Female
At Rest	In Motion
Straight	Curved
Light	Darkness
Good	Evil
Square	Oblong

Until only recently, our modern scientific view of the universe hinged on the existence of one other antithetical pair: particle versus wave. The former is best exemplified by a pebble, the latter by ripples on water.

In certain crucial ways, the two play by very different rules.

A pebble has a well-defined shape and size. By contrast, ripples are smeared out, like the widening circular swells caused when we throw something into a pond.

A pebble exists autonomously and identically in any medium; its shape and size is the same in water or in molasses. By contrast, the substance and appearance of ripples depend critically on whether they are moving in water, molasses, or some other material. To borrow a phrase from Marshall McLuhan, when it comes to waves, the medium *is* quite literally the message.

For centuries following the ancient Greeks, scientific-minded scholars stubbornly upheld the belief that a particle is a particle and a wave is a wave. For instance, a planet is a particle, period — it can never behave like an ocean wave. Likewise, light is a wave and can never be like an atom, which is clearly a particle. As Rudyard Kipling wrote in *The Ballad of East and West* to describe the very different cultures of India and Britain:

> Oh, East is East, and West is West, and never
> the twain shall meet,
> Till Earth and Sky stand presently at God's
> great Judgment Seat.

I cannot overstate how incredibly certain late nineteenth-century scientists were about the fixed and separate identities of wave and particle — about our alleged understanding of *everything*, actually. In 1900 the eminent Scottish physicist William Thomson — vastly more celebrated in his day than Stephen Hawking is today — is said to have boasted, "There is nothing new to be discovered in physics now. All that remains is more and more precise measurement."[2]

But Thomson's rosy picture of things was soon shattered. During the early years of the twentieth century, scientists were rocked back on their heels when they discovered how wrong they had been about particles and waves. They woke from their ignorance, like Dorothy after the tornado, to a fantastic new world stranger even than Oz.

The paradigm-shifting discovery happened when science took a close look at the behavior of atoms and led to what we now call quantum theory. As Niels Bohr, responsible for many of the stunning revelations, reportedly put it, "If anybody says he can think about quantum theory without getting giddy, it merely shows that he hasn't understood the first thing about it."[3]

Particles? Waves? Never the twain shall meet? Hold the phone. According to quantum theory, our age-old beliefs fail to describe the actual complex, contradictory nature of the atomic world, of the now particle-like, now wave-like pixels of physical reality.

The experiment that first revealed the odd nature of atoms was performed in the late 1800s, first by Heinrich Hertz and then by his assistant, Philipp Lenard. The setup was pretty simple. When the scientists shone light of different colors and levels of brightness onto a polished metal plate, their instruments detected negatively charged particles — electrons, we later called them — coming off the surface. The men reasoned that electrons in the metal plate were being shoved around by the light waves the way surfers are by ocean waves. Nothing unusual about that.

But something wasn't quite right. Instead of appearing to be nudged off the metal plate by soft-body waves, the electrons veritably leaped from the surface as if hit by a heavy sledgehammer.

It reminds me of an extraordinary phenomenon I discovered at the Avalon Peninsula in southeast Newfoundland. While hiking there, my ears picked up on a strange sound. I headed in that direction, and minutes later I came upon a beach made entirely of large pebbles. Every time a wave broke on shore, it jostled the pebbles, creating a loud racket. The pebbles on that beach behaved exactly the way science had expected electrons would behave in Heinrich Hertz's experiment. The electrons, according to the old theory, should have been jostled around. But instead, they flew off the metal surface like sharply struck billiard balls.

There was another bit of unexpected strangeness. The intensity of the light source made no difference. When hit with dim or bright light, electrons flew off the plate with the same energy, which made no sense. It was like saying that tiny ripples and monster waves propel surfers with equal momentum.

And that wasn't the end of it. There was one final bit of nuttiness. Red light pushed electrons off the plate with far less energy than blue light. Color corresponds to a light wave's *frequency* — the number of cycles per second it has — which should have nothing to do with the wallop it conveys to an electron. In short, everything about the experimental results was backward. The *brightness* of a light wave, which should determine its wallop, didn't. And the *color* of a light wave, which shouldn't make any difference, did.

In 1905 Albert Einstein was the first to explain Hertz's so-called photoelectric experiment's crazy results. He was only twenty-six years old, a total newcomer to the world of professional science; yet, two decades later he would be awarded the Nobel Prize in physics for his historic achievement.

What Einstein explained was this: light waves don't always behave like waves; sometimes they behave exactly like particles. These queer Jekyll-Hyde-like particle-waves — light quanta, as they came to be called — have an energy that corresponds to what we call *color*. Their total population tallies with what we call *brightness*.

Here's how Einstein put it: "The energy of a light ray spreading out from a point is not continuously distributed over an increasing space [i.e., it doesn't behave like a wave], but consists of a finite number of energy quanta [i.e., simple particle-waves] which are localized at points in space, which move without dividing, and which can only be produced and absorbed as complete units."[4]

This pronouncement defied everything modern science had thought was beyond questioning. It was as if Einstein were declaring that odd could be even, right could be left, white could be black. In the history of modern science, no one had ever spoken such seeming contradictory nonsense.

As if that weren't enough of a blow for early twentieth-century science, there soon appeared a French aristocrat — like Einstein, a parvenu — who completed the job of destroying the cherished particle-wave dichotomy. His name was Prince Louis-Victor de Broglie. In 1924, at the age of thirty-two, young de Broglie floated the idea that since waves can behave like particles, then surely particles can behave like waves. "After long reflection in solitude and meditation," he recounted later, "I suddenly had the idea, during the year 1923, that the discovery made by Einstein in 1905 should be generalized by extending it to all material particles and notably to electrons."[5] The young prince turned out to be correct, which led to his being awarded the 1929 Nobel Prize in physics.

Today undergraduates — even high school kids — can conduct a simple experiment that illustrates de Broglie's unprecedented assertion. All it takes is an electron gun (like the kind at the back of an old-fashioned TV tube), a thin slice of carbon, and a detection screen of some sort (like the luminescent face of a traditional TV set).

High-speed electrons passing through the carbon crystals emerge from it and create a scatter-shot pattern on the screen that is identical to the diffraction pattern we get from shining light waves through a crystal. It is not at all the outcome we would observe if the electrons were simply bullet-like particles.

Particles behaving like waves, waves behaving like particles. If it all sounds a bit confusing to you, welcome to the dizzying world of twenty-first-century science. It's a world that sounds senseless but appears to describe very well the universe in which we live, a universe that as far as we know comprises objective truth. A world where, in certain circumstances, light behaves like a wave and in others like a particle. Ditto for an electron; sometimes it behaves like a particle, sometimes a wave.

Please understand, it is not a world in which real objects are half one thing and half a completely opposite thing. What science appears to have discovered is that a light ray and an electron are each *fully* a wave and *fully* a particle. Each is the embodiment of a contradiction.

It's a world where our language has met its match. A world where particles and waves are all one hard-to-name, hard-to-comprehend thing — objects with a kind of dissociated identity, whose weird behavior is usually evident only in the teeny-tiny realm of atoms, but not always. (For more on this idea, see chapter 9.) Bizarre, seemingly contradictory objects of which we ourselves are made.

The Bible

The gospel of Mark recounts that "Jesus and his disciples went on to the villages around Caesarea Philippi." On the way, Jesus asks them, "Who do people say I am?" (8:27).

Such a simple question. Yet no historical person's identity — not Buddha's, not Moses', not Muhammad's — has ever stirred up so much emotion or had such a profound and lasting influence on the world.

I AM WHO I AM

Jesus asked, "Who do people say I am?"

On the one hand, he was a seeming nobody, ostensibly a carpenter, an itinerant preacher, a soon-to-be-convicted Jew hailing from a remote, maligned part of Israel. But on the other, at the very least, he was the founder of what has become the world's largest, and by far most widespread, religion. Talk about a contradiction.

Jesus' message is one of love, peace, forgiveness, selflessness, humility, and nonmaterialism, yet to this day — two thousand years after his time on earth — his very name rouses feelings of anger and hatred in many people. That is yet another glaring contradiction.

Louis Lapides is the founding pastor of Beth Ariel, a messianic congregation in Sherman Oaks, California. As a young Jew, before his conversion to Christianity, Lapides recalls that neither he nor his parents nor anyone he knew paid any real attention to Jesus. "Basically, he was never discussed." As for showing respect — well, Lapides says, if the name Jesus was ever mentioned, it was "only derogatorily."[6]

Lapides's personal recollections jibe with the experiences of my own messianic Jewish friends. From them I hear heartbreaking stories about having been shunned by relatives — even their own parents — who consider their conversions somehow treasonous. Their family members despise the cross and consider the Christian scriptures to be a fairy tale at best.

Lapides himself recalls, "When the New Testament was first presented to me, I sincerely thought it was going to be a handbook on anti-Semitism: how to hate Jews, how to kill Jews, how to massacre them."[7]

Jesus asked: "Who do people say I am?"

These days we can come across just about any answer imaginable by surfing websites or watching movies such as *The Last Temptation of Christ* and *The Da Vinci Code*. According to such sources, Jesus was a megalomaniac, a criminal, a conspirator, a closeted gay, a womanizer, a married man, a trickster, and on and on.

Historically, the answers most people have given to Jesus' pointed question fall into two broad, opposite categories — the analogs to science's particle and wave dichotomy. One category holds that Jesus was nothing more than a man; the other insists he was God.

What are the fundamental differences between the two? It is hard to be precise and, I suspect, impossible for all of us to agree on any single answer. But let me take a stab at it. As I see it, God is unadulterated spirit, infinite in space, time, knowledge, power, and creativity, and he is unchanging. By contrast, a man is incarcerated by the material world, finite in space, time, knowledge, power, and creativity. And he always is changing.

Which of these two classes of being do you think accurately describes Jesus?

Officially, Jews see Jesus as a man — on the negative side, a convicted heretic and all-around troublemaker; and on the positive side, a popular teacher, reformer, ethical master, even saint. But definitely not the Messiah. Muslims, too, see Jesus as a man, a first-century prophet who ultimately was trumped by the seventh-century prophet Muhammad. Buddhists see Jesus as a man as well, a wise, enlightened man, to be sure, but a man nonetheless. Hindus vary in their opinions. Some see Jesus as a god (though not the incarnation of the one and only God), others as an exemplary man. "To me," Mahatma Gandhi wrote in *What Jesus Means to Me*, "he was one of the greatest teachers humanity has ever had."[8]

Fourth-century Gnostics saw Jesus as a noneternal spiritual creature but not as God. Today's Jehovah's Witnesses believe that Jesus is God's first creation but not God himself. A small sect calling themselves Christadelphians claim that "Jesus is a man, who was tried and tempted as we are, yet who resisted sin even till death."[9]

In AD 325, Constantine the Great — a Christian convert who credited Jesus for military victories leading to his emperorship — convened a council of Christian bishops hailing from all over the known world. Constantine wanted the bishops to confront head-on the various wide-ranging estimations of Jesus (especially the one being aggressively promoted at the time by Arius, a Gnostic). He wanted the council to reach a consensus by arriving at the most discerning conclusion that biblical and historical evidence could support.

Read this fascinating eyewitness account of the gathering recorded by Eusebius Pamphilus, the Greek-born bishop of Caesarea: "There were more than 300 bishops, while the number of elders, deacons and the like was almost incalculable. Some of these ministers of God were eminent for their wisdom, some for the strict living, and patient endurance of persecution, and others for all three.... The Emperor provided them all with plenty of food."[10]

Certain critics of the First Council of Nicaea are wont to complain that the bishops used it somehow to hijack Christianity, to twist it into something it wasn't really. As a physicist, I see it very differently, as more akin to the process science uses constantly to reach a consensus about what is credible and what is not, based on the best available evidence.

At the conclusion of the First Council of Nicaea, the bishops reached a consensus concerning a number of important matters, including when exactly to celebrate Easter. But above all, they reached consensus on the definitive answer to the question Jesus had put to his disciples: "Who do people say I am?"

In the words of the Nicene Creed, which to this day is the Christian equivalent of a scientific truth, the answer is this: Jesus, you are "the only-begotten Son of God, begotten of the Father before all worlds; God of God, Light of Light, very God of very God; begotten, not made, being of one substance with the Father, by whom all things were made." In other words, Jesus is both human *and* God.

"I and the Father are one," Jesus himself states plainly in John 10:30. Which is why John 14:6 makes perfect sense when Jesus explains, "I am the way and the truth and the life. No one comes to the Father except through me." If you reject Jesus, you reject God himself, because the two are one and the same.

Admittedly, the radical notion that Jesus is simultaneously God *and* man is as confounding as the quantum theoretical idea that light is at once wave *and* particle, and an electron is concurrently particle *and* wave. Not half 'n' half, please note, but *fully* one and *fully* the other.

To me, Jesus best reveals himself to be fully God in the unique way he forgives. I really like the way Dr. Donald Carson, professor of New Testament at Trinity Evangelical Divinity School, explains it. Carson, interviewed in Lee Strobel's excellent book *The Case for Christmas*, says, "The point is, if you do something against me, I have the right to forgive you. However, if you do something against me and somebody else comes along and says, 'I forgive you,' what kind of cheek is that? The only person who can say that sort of thing meaningfully is God himself, because sin, even if it is against other people, is first and foremost a defiance of God and his laws."[11]

It is precisely why, as recounted in the gospel of Mark, Jesus created such a huge commotion when he said to the paralyzed man, "Son, your sins are forgiven." Teachers of the law, standing nearby, flew into a rage: "Why does this fellow talk like that? He's blaspheming! Who can forgive sins but God alone?" (2:5–7).

Who, indeed.

On the flip side, for me, Jesus best reveals himself to be fully a man through his emotionality. As when he heard about the death of his friend Lazarus. In John 11:35, we read that "Jesus wept." We see his human emotions coming to the fore again when contending with God the Father in the garden of Gethsemane. Jesus never sinned, but he came pretty close to behaving disobediently when he begged God to let him off the hook. "'*Abba*, Father,' he said, 'everything is possible for you. Take this cup from me'" (Mark 14:36).

In both situations, Jesus behaved exactly as I, a mere human, would have. I would have wept and waffled. And because of that — because of the Messiah's miraculous "dissociative identity" — the God of Israel, who was once but a disembodied voice, a burning bush, a pillar of cloud by day and a pillar of fire by night — that self-same God is now someone I can both fully worship and fully befriend.

What Does It All Mean to You and Me?

One of the practical lessons from this discussion, I believe, is this: at the extremities of reality — in the realms of the super-tiny and the supernatural — we come face-to-face with truths for which we have no adequate words. Foolish-sounding truths that are paradoxical through-and-through.

These contradictory truths open our eyes to deep mysteries about God's creation that don't fall neatly into our conventional ways of thinking, which should inspire a healthy humility in us. They affirm the words in 1 Corinthians 1:20, 27: "Where is the wise person? Where is the teacher of the law? Where is the philosopher of this age? Has not God made foolish the wisdom of the world?... God chose the foolish things of the world to shame the wise."

This brings us to the next lesson: you and I should take great care not to be fooled by seemingly foolish appearances. Most great truths and most real geniuses were scorned at first.

For that reason, you and I mustn't judge too quickly possible sources of illumination. Before Christianity, before quantum theory, who would have guessed that amazing truths about the universe would appear to us cloaked as crazy-sounding contradictions?

I am reminded of the conversation between young Alice and the wizened Queen in Lewis Carroll's classic tale *Through the Looking Glass*:

> Alice: "There's no use trying, one *can't* believe impossible things."
>
> Queen: "I daresay you haven't had much practice. When I was your age, I always did it for half-an-hour a day. Why, sometimes I've believed as many as six impossible things before breakfast."[12]

As the topsy-turvy weirdness of quantum theory was being revealed, leaving him to wonder what to expect next, physicist Niels Bohr reportedly said, "The opposite of a correct statement is a false statement. But the opposite of a profound truth may well be another profound truth."[13]

The final lesson for you and me, I believe, is this: like Jesus, like a light ray, like an electron, we ourselves are living, breathing contradictions. That is what Robert Louis Stevenson was saying in *Dr. Jekyll and Mr. Hyde*. Listen to Dr. Jekyll musing on his plight: "It was on the moral side, and in my own person, that I learned to recognize the thorough and primitive duality of man; I saw that, of the two natures that contended in the field of my consciousness, even if I could rightly be said to be either, it was only because I was radically both."[14]

The Bible tells us that we are both flesh and spirit, and the two are always contending. In fact, the whole of the Scriptures is a historical account of our inner struggle between love and hate, light and darkness, arrogance and humility, knowledge and wisdom, pleasure and joy, honesty and deceit, sinfulness and innocence, enmity and forgiveness, obedience and rebellion, selfishness and sacrifice. And, in the end, life and death.

Louis Lapides knows all about the war between spirit and flesh, between our own Jekylls and Hydes. He recalls that as a young Jew, he experienced very little of the Spirit; he felt disconnected from God. "In Judaism I didn't feel as if I had a personal relationship with God. I had a lot of beautiful ceremonies and traditions, but he was the distant and detached God of Mount Sinai, who said, 'Here are the rules — you live by them, you'll be okay; I'll see you later.'"[15]

Lapides's life took a nose dive after his parents divorced and he served time in Vietnam. He surrendered completely to the flesh, to the excesses and infatuations of the 1960s — among them drugs, rebellion, and Eastern mysticism, which taught him that he himself was God. At one point, he even befriended Satan worshipers and accompanied them to meetings. Around that same time, he also contemplated suicide. "I guess I accepted Satan's existence," he recalls, "before I accepted God's."[16]

The story of Lapides's conversion, as recounted in *The Case for Christ*, is well worth reading. For now it is enough to know how he explained his conversion to family and friends who were incredulous at the sudden changes in him. "Well, I can't explain what happened," he told them truthfully. "All I know is that there's someone in my life, and it's someone who's holy, who's righteous, who's a source of positive thoughts about life — and I just feel whole."[17]

The two words in Lapides's testimony that leap out to me are "someone" and "whole." For Lapides, God went from being an abstraction to *someone* dwelling within him — someone fully God and fully human. In the end, Jesus mended his fractured, tormented life — his dissociated identity disorder, if you will — and made him *whole*. Jesus, himself the perfect marriage of spirit and flesh, united Lapides's own warring spirit and flesh.

That, I believe, is the biggest lesson and greatest power of science's and the Bible's shared belief in the possible reconciliation of seemingly antipodal truths. *Knowing* about the possibility changes our entire understanding of the universe and of God. *Living* the possibility changes us — from creatures that are half Jekyll and half Hyde to ones that are fully what God intends us to be.

CHAPTER 4

SEEING IN THE DARK

SIGNIFICANT PARTS OF REALITY ARE HIDDEN FROM US

*So we fix our eyes not on what is seen, but
on what is unseen, since what is seen is
temporary, but what is unseen is eternal.*

2 CORINTHIANS 4:18

*Equipped with his five senses, man
explores the universe around him
and calls the adventure Science.*

EDWIN HUBBLE

SEEING IS BELIEVING, RIGHT?

Before you answer too quickly, ask yourself this: Is it really true that we believe in only those things we actually have seen with our own eyes? Of course it isn't.

I remember once getting to see a rare "ghost lizard" — an Ozark grotto salamander — deep inside Meramec Caverns in southern Missouri. Its skin was wan and its eyes were fused shut. Why? Because without light, eyes are useless. Without light, there is no sight.

During the day, human eyes depend primarily on the photoreceptors in the retina called cones; at night, the photoreceptors called rods. The handoff between the two — between our retina's day shift and night shift — starts around dawn or dusk and is completed by the time the ambient light level has reached that of dim moonlight. Cones and rods complement each other in other ways as well.

I remember doing a *Good Morning America* report on a young man named Dave Weber, who was born with a rare condition called achromatopsia. Dave's retinas lack cones, which means he doesn't see well in daylight. It also means he is completely blind to colors. For him, the universe is a billion shades of gray. "If everyone had conspired to tell me the sky was purple," he quipped during my interview with him, "I would still believe it. I wouldn't know any differently."[1]

But Dave's condition has an upside. His rods are so sensitive to light that as soon as the sun starts to set, the world comes alive for him. He says: "All of the things that were kind of washed out during the day start to become very clear and very crisp, and very, you know, very detailed."[2]

Some people with achromatopsia are able to read a license plate four blocks away in the dark! This superhuman ability is of interest to military engineers developing night-vision cameras.

Science and the Bible, I believe, offer us a similar kind of superhero-like ability to peer into the darkness of human ignorance. As we are about to see, they are both time-honored and time-tested resources that enable us to perceive objective truths hidden deep within the shadows of the remotest regions of the physical universe *and* the spiritual one as well.

Science

Seeing in the dark is not easy. Science knows that very well.

For centuries chemists were in the dark about fire. They wondered, *What makes some materials combustible and others not, and why does a fire eventually burn itself out?* As they groped about for answers, they came to believe strongly in the existence of a colorless, odorless, tasteless, massless flammable substance called *phlogiston*.

Flammable materials allegedly contained lots of it, others not so much. In any case, the argument went, fire fed on this phantom phlogiston, so when it was fully consumed, a fire would naturally go out.

In the late eighteenth century, however, Joseph Priestley and Antoine-Laurent Lavoisier discovered oxygen and the true composition of air. Suddenly, scientists realized it was oxygen that fed fires; phlogiston was just a hallucination.

A similar thing happened to nineteenth-century physicists. They were in the dark about starlight, wondering aloud, *How can light possibly travel across vast stretches of nothingness to reach earth?*

At the time, you see, physicists believed that light consisted of waves, and that waves needed a medium through which to propagate. Ocean waves needed water; sound waves needed air (or some other substantial medium). Surely, physicists reasoned, light waves required some kind of ubiquitous substrate to enable their movement.

Eventually scientists decided that empty space must be filled with a colorless, odorless, tasteless, massless substance called *ether*, a diaphanous causeway that enabled starlight to reach us over immense distances. But by the early 1900s, after a long string of experiments were conducted — most notably one by American scientists Albert Michelson and Edward Morley — nary a trace of the alleged ether could be found. Physicists were forced to concede it had been only a hallucination.

We must be careful not to be overly harsh on science for its occasional fantasies. After all, it spends a lot of time in the dark. That is its purpose: to help us explore and shed light on aspects of the physical universe we don't understand.

I like the way British physicist Brian Cox puts it: "There are places out there, billions of places out there, that we know nothing about. And the fact that we know nothing about them excites me, and I want to go out and find out about them. And that's what science is."[3]

The fact that science is still in business is a consequence of its considerable success, but also its never-ending failures. Never-ending because of this truth: at any given time, significant parts of reality are hidden from us — so hidden that even the vaunted methods of science are hard-pressed to drag them into the light.

These days, among the various scientific disciplines, it is astrophysics that is very much in the dark about things, especially gravity. Gravity is everywhere, so you'd think we'd know all about it. We don't.

How mysterious is it? Let me count the ways.

Gravity is unusual because, unlike other forces, it is a universal property of matter. The electric force, for example, is generated only by electrically charged particles, not neutral ones. And whereas the electric force comes in two varieties — attractive and repulsive — gravity is always attractive. There is no such thing as repulsive gravity, or antigravity, except in science fiction. Even the gravitational force of antimatter is attractive.

Gravity is also mysterious because, while it is the weakest of all known forces — considerably weaker than atomic and nuclear forces — it is the most indomitable and far-ranging. Our bodies generate gravitational forces that cannot be stopped — that extend clear across the universe — which is why no one has ever been able to invent a gravity shield. As the famous physicist Paul Dirac reportedly observed, "Pick a flower on Earth and you move the farthest star."[4]

Sir Isaac Newton was the first person to describe accurately *how* gravity behaves. These days every high school student learns about his famous law of gravity:

$$F = GMm/r^2$$

But even Sir Isaac couldn't make heads or tails of *what* gravity is, really. "Gravity must be caused by an agent, acting constantly according to certain laws," he ventured, "but whether this agent be material or immaterial, I have left to the consideration of my readers."[5] In other words, he was saying, "Hey, your guess is as good as mine."

In 1915 Albert Einstein offered a fantastical explanation of gravity called the general theory of relativity. I dare say that never has so much arcane mathematics been used to describe something so seemingly commonplace. Here is Einstein's makeover of Newton's simple law of gravity, which includes a revision he published in 1917:

$$G_{\mu\nu} + g_{\mu\nu}\Lambda = \frac{8\pi G}{c^4} T_{\mu\nu}$$

I took my first course in general relativity during my second year of graduate studies at Cornell. Our textbook — a legendary, black-clad tome titled simply *Gravitation*[6] — was thicker than the Manhattan phonebook. You can buy it on Amazon; it is 1,215 pages long!

Fortunately, the gist of Einstein's explanation of gravity is not hard to comprehend. Picture yourself sitting alone on a large, soft-cushioned couch. That image represents you stationed in outer space. Now suppose a huge football player sits down next to you. What happens? He creates a sprawling concavity in the couch that causes you to slide toward him. That is you in outer space being attracted gravitationally to a nearby giant planet.

According to general relativity, massive objects create large, bowl-shaped depressions in space-time — which, in turn, cause everything around to fall into them. That's gravity. The more massive an object, the deeper the basin, the more powerful the attracting force.

Earth and the solar system's seven other planets are drawn toward the sun — they orbit it constantly — because the sun creates a cavernous depression in space-time. How cavernous? It's some *five billion* miles across and then some. (Remember what I said: gravity extends infinitely far. Like old soldiers, you might say, it fades away but never dies.)

The solar system, in turn, is drawn toward the center of the Milky Way galaxy — orbits it constantly — because the galaxy creates a space-time basin even bigger than the sun. Roughly 600,000,000,000,000,000 (600 quadrillion) miles across! From there it levels off gradually in all directions.

And so it goes throughout the universe.

The general theory of relativity was initially neither well understood nor warmly received, partly because the mathematics Einstein used to describe gravity was devilishly complicated. It didn't help the theory's credibility either, when, just a year after its debut, German astrophysicist Karl Schwarzschild discovered that Einstein's newfangled law of gravity predicted the existence of celestial ghosts — massive objects that could not, *by definition*, be seen.

Could this be just another one of science's nighttime deliriums, doomed to be debunked by the light of reason? In the minds of scientists who cherished Newton's old, simple law of gravity, general relativity circa 1916 seemed to be in danger of going the way of phlogiston and the notorious ether.

In 1919, however, British astronomer Sir Arthur Eddington performed the first real experimental test of Einstein's new theory, and suddenly everyone saw the light. Literally. Eddington began by noting what all astrophysicists at the time knew, which is that the path of a light ray passing close to the sun would be drawn toward it by gravity, bent by a measurable amount, no differently than a planet. But Einstein's equation predicted *twice* the amount of bending that Newton's equation did. So here was a perfect opportunity to see which theory was right.

On May 29, on the island of Príncipe (off the west coast of Africa), Eddington and his people watched expectantly as two things happened simultaneously: (a) the sun went into a total eclipse and (b) the sun's shrouded disk slowly passed across the bright star cluster Hyades. This marvelous coincidence spared Eddington from being blinded by sunlight as he watched what happened to the starlight. He could measure precisely how much the starlight's path was bent.

What did he find out? Eddington determined that general relativity's prediction, not Newton's, was right on the money. Einstein was vindicated.

General relativity scored a second decisive victory in 1929, when American astronomer Edwin Hubble discovered that the universe was expanding. Galaxies were rushing away from one another at a considerable rate. This was shocking news, wholly unexpected — even by Einstein.

Like everyone in science at the time, the German genius had always assumed that the universe was static. (For more on this idea, see chapter 6.) When formulating general relativity, therefore, it had concerned him that the equations predicted a universe that was *not* static. To correct this seemingly appalling flaw, he had added a fudge factor (a repulsive force that counteracted gravity's attractive force) to make things come out static. When he learned of Hubble's discovery, he kicked himself for having done this.

Astrophysicist Mario Livio, author of *Brilliant Blunders*, says: "He [Einstein] definitely regretted it. He wrote about that to a number of friends. He thought it was ugly."[7] Fortunately, it was an easy fix: Einstein merely retracted the fudge factor and, *voila!*, general relativity was in line perfectly with Hubble's history-making discovery.

General relativity's two experimental bull's-eyes turned the wiry-haired, pipe-smoking Einstein into a public sensation, the Babe Ruth of physics. It also forced astrophysicists to take seriously Schwarzschild's ghostly objects — *black holes*, as they came to be called. Today, a century later, astrophysicists claim to detect the unseen presence of black holes practically every day. Black holes have become ho-hum — if that's possible — and general relativity remains securely seated on the throne of scientific opinion.

But the enormous darkness surrounding the subject of gravity has not been dispersed. In the years since 1915, we have learned that black holes are but one tiny example of gravity's most deeply hidden secrets — and not even the most vexing. That superlative distinction belongs to a mystery I first heard about as a freshman grad student. Back then it was called the "missing mass" problem.

As a rule of thumb, a galaxy's rate of spinning ought to be proportional to its total mass. But in the early 1930s, Dutch astronomer Jan Oort discovered that a typical galaxy spun far faster than that.

Swiss astronomer Fritz Zwicky observed the same problem with a typical *swarm* of galaxies: they whirled around their common center way faster than their apparent masses justified. Worse, it wasn't a small discrepancy. Zwicky's calculation indicated that to account for the unexpectedly rapid spinning, the objects should be about ten times as massive as they appeared to be. *Ten times!*

The implication of Zwicky's revelation was that either (a) Einstein's theory of gravity was wrong, or (b) a good deal of a galaxy's or galaxy cluster's mass was unseen by us. It wasn't a pretty choice.

Today astronomers have come down on the side of believing that galaxies and galaxy clusters are pregnant with some sort of exotic material that is invisible to us — they are calling it *dark matter*. They haven't yet identified what it is exactly — or even established that it truly exists — but it is not for lack of trying. Despite decades of using every imaginable means of detection — from gamma-ray telescopes in outer space to cryogenic subatomic particle monitors buried deep inside a northern Minnesota mine — their occasional, tantalizing reports of success remain as unreliable as Elvis sightings.

And dark matter isn't even the most astonishing thing modern astrophysicists have discovered about gravity. That supreme credit is reserved for what happened in 1998. I was at ABC News at the time and covered the story when it broke. I will never forget it.

On December 17 a team of astrophysicists announced that in their years-long observations of supernovae (exploding stars, some of which end up being black holes), they'd found evidence that the universe was not only expanding, but *accelerating*. This was alarming news because, ever since Hubble's discovery, there had been every indication that the expansion of the universe was gradually slowing down — which made sense. Little by little, gravity's indomitable attractiveness was putting the brakes on the explosive, inflationary aftermath of the hypothesized big bang.

So why in the world was the universe *accelerating*?

Today astrophysicists are still in the dark about how to answer that very pressing question. But they have agreed on a name for whatever material agent is behind the acceleration. They are calling it *dark energy*. Some astrophysicists suggest that dark energy is a property of space-time itself, that ironically Einstein's infamous fudge factor is needed after all because it represents a repulsive force, just the thing to cause the acceleration. Others speculate that dark energy is a new twist on the old, discredited ether; an omnipresent, repulsive material many are calling the *quintessence*. Others still are betting that dark energy is related somehow to the quantum vacuum, whose own weirdness makes black holes seem as ordinary as watermelons. (For more on this idea, see chapter 6.)

All told, astronomers have concluded that dark energy comprises some 68 percent of the total universe and dark matter, about 27 percent. That means only 5 percent of the entire universe is visible to us![8] That astonishing revelation bears emphasizing. Everything we call scientific knowledge is based on but a pittance of what there is to know about our world. *Ninety-five percent* of it is hidden from us.

In light of this latest bombshell, do we stand a chance of ever really understanding gravity? Astronomers are hard at work believing they can. But they must labor with the unsettling awareness that our science is 95 percent in the dark about the universe it seeks and claims to understand; about what is real or not, what is possible or not — even about a prosaic force that exists literally right under our noses.

The Bible

Death is no less common, no less familiar to us than gravity. Yet on any list of things that keep us in the dark and really matter to us, I think it is safe to say that death ranks much higher than gravity. It most likely ranks second only to the mysteries of life and God.

We wonder, *Is death a permanent, moribund state? Or is it, like birth, a remarkable metamorphosis, resulting in an altered form of life?*

"I believe that when I die I shall rot," said Bertrand Russell, the noted British mathematician.[9]

Stephen Hawking, who has a simple, mechanical view of life and death, agrees: "I regard the brain as a computer which will stop working when its components fail. There is no heaven or afterlife for broken-down computers; that is a fairy story for people afraid of the dark."[10]

American actor Joaquin Phoenix is just as sure of it: "I don't believe in god. I don't believe in an afterlife. I don't believe in soul. I don't believe in anything."[11]

The opinions of such well-known people naturally attract a great deal of attention. But in the grand and diverse scheme of human beliefs, they uphold a minority view of what happens to us when we die.

In late 2011, Reuters and the Ipsos Social Research Institute conducted a poll of 18,829 people from twenty-three nations. Only 23 percent of respondents expressed the belief that death means we just "cease to exist." Everyone else said they are certain (51 percent) or not sure (26 percent) that there is life after death.[12]

Believing in an afterlife is consistent with believing there is a huge part of reality that is hidden from us. *Not* believing in an afterlife is in line with believing that what we perceive, what we experience in the here and now, is all there is. But as we have just seen, science itself has discovered that there is a great deal about the "here and now" we can't see and certainly don't understand. Most of it — black holes, dark matter, dark energy, and who knows what else — is no less incredible than the claim of an afterlife.

Christianity — now representing roughly one-third of the world's population — has always been clear about its position on the afterlife. Jesus conquered the grave, and so will we one day, for better or worse. "Why do you look for the living among the dead?" the angels said to the women seeking Jesus in the tomb. "He is not here; he has risen!" (Luke 24:5–6).

Christianity's view of the afterlife is quite fantastical compared to that of any other major religion. In fact, it wouldn't be far off the mark to say that Christianity shook up religion's traditional beliefs about an afterlife the way Einstein shook up science's traditional view of gravity.

Christianity accomplished this in at least two ways. Historically, and to this day, most afterlife beliefs have centered on a meritocracy of some sort. It is true even in Judaism, Christianity's progenitor, which generally pays very little attention to the subject. "Reform Judaism," explains Rabbi Howard Jaffe, "while not taking any 'official' position on the matter, has for the most part ignored the question, and tended towards the belief that there is no such thing."[13]

To the degree that an afterlife is part of Judaism's conversation, it is seen as something we must earn. Our good deeds, our *mitzvoth*, are what win God's favor and get us into *Olam Ha-Ba*, the next world.

But the motivation for doing good deeds should not be the desire to get into the next world. "The classic text regarding this matter is paraphrased as follows," explains Rabbi Jaffe: "Do not be as ones who labor for their master mindful of the reward that will be coming, but rather as those who serve their master with love and with joy."[14]

Buddhism and Hinduism — Eastern religions generally — differ from Judaism in many significant ways. But in this one respect, at least, they are like-minded. It is the idea that if there is any kind of afterlife *and* if it offers the possibility of pleasure, then the reward must be hard-won. Perhaps it is one reason why so many apostate Jews feel comfortable becoming Buddhists.

Summed up, the meritocratic view of the afterlife is this: God doesn't love us automatically and unconditionally; his love must be deserved somehow. It is a very human idea — that there is no free lunch — but one that Jesus squarely took on and radically reformed with his life, death, and resurrection.

Jesus brought to earth the message that the afterlife is a *choice*, not a chore. That it is radically inclusive, not exclusive, as some misperceive it to be. For Christians this is a fundamental truth of the universe — that heaven awaits anyone and everyone who simply and sincerely accepts God's gift of forgiveness. Good deeds — the *mitzvoth* — proceed from such a choice; they don't precede or outrank it.

There is a second way that Christianity radicalized our religious beliefs about the afterlife.

Many religions — and even a-religious peoples — tend to believe that if there is an afterlife, it involves our souls or spirits, not our bodies. It is a view that arises naturally from dualism, the hoary conviction that material reality and spiritual reality are antithetical — the former is temporal and evil; the latter, eternal and thoroughly good. When we die, dualists believe, our bodies are discarded and left behind; our souls survive and enter a spirit world, or set up shop in other bodies or entities.

"Persistence of consciousness" — that is what this anything-but-new belief is called by New Agers such as Deepak Chopra, author of *Life after Death: The Burden of Proof*. In a published debate with someone skeptical about an afterlife, Chopra wrote this: "It's a shame that he doesn't grasp that the afterlife is about nothing but consciousness. (I don't offhand know anyone who took their bodies with them.)"[15]

But a bodily resurrection is precisely what Christianity teaches. Radical? You bet. Christianity's belief is based on the Bible's historical account of Jesus' bodily resurrection. Also, it is memorialized in the Apostles' Creed ("I believe in ... the resurrection of the body"), the definitive statement of faith credited to the twelve disciples, who knew Jesus personally and witnessed his ministry firsthand.

I'm reminded of my first overseas assignment as a television reporter. The year was 1988. I flew to Cairo to cover the unveiling of the 3,300-year-old mummified remains of Ramses II, the legendary pharaoh who claimed to be a god.

Had Jesus experienced a merely *spiritual* resurrection, the kind touted by dualist religions, his dead body would have stayed behind — and possibly been mummified and put on display à la the faux deity Ramses II. His tomb would not have been empty. Easter wouldn't exist to celebrate a risen Christ.

But it does exist, and the very thing that happened to the crucified Jesus, Christianity believes, will one day happen to us: "Listen, I tell you a mystery: We will not all sleep, but we will all be changed — in a flash, in the twinkling of an eye, at the last trumpet. For the trumpet will sound, the dead will be raised imperishable, and we will be changed" (1 Cor. 15:51–52). According to Paul, the writer of the foregoing passage, when we die, we become spiritual beings that somehow remain material. I like how Paul doesn't pretend to understand it. Credit him for admitting the deep mystery of the bodily resurrection.

Happily, in other passages, the Bible offers clues about the nature of this altered form of life. We know it is *material* because the resurrected Christ walked, talked, spoke, ate, and drank as if he were still human. He even allowed Thomas to inspect his wounds. "Put your finger here," he said to the skeptical disciple; "see my hands. Reach out your hand and put it into my side. Stop doubting and believe" (John 20:27).

But we also know our resurrected bodies will be *nonmaterial.* They will not be subject to the laws of physics, as seen in the way the risen Jesus was able to get around seemingly instantaneously and without any hindrance. Here is John's uncanny report of how Jesus materialized to the disciples just before the famous encounter with Thomas: "A week later his disciples were in the house again, and Thomas was with them. Though the doors were locked, Jesus came and stood among them and said, 'Peace be with you!'" (John 20:26).

This passage seems to indicate that when we die, each of us will retain our current identity but will inhabit a new body. This surmise is borne out by 1 Corinthians 15:42–44, which declares that whereas our current form "is perishable, it is raised imperishable; it is sown in dishonor, it is raised in glory; it is sown in weakness, it is raised in power; it is sown a natural body, it is raised a spiritual body." In short, according to Christianity, our resurrected bodies will be made of a substance that is truly exotic — as different as, say, dark matter or dark energy is from visible matter or prosaic energy, or anything else we currently understand.

Unbelievable? To some, clearly, yes — especially to the 23 percent who believe we will simply cease to exist when we die. But it shouldn't be that unbelievable, really — not to a species accustomed to grappling intelligently with the great mysteries of the universe.

Like, say, gravity.

What Does It All Mean to You and Me?

There is more to us and life than meets the eye, materially and spiritually — so we should live like it. I believe that is the first and most obvious practical lesson from this discussion.

Both science and the Bible agree that our existence is influenced by information that is almost *all* hidden from our worldly perspective. Our existence is steered to a large extent by forces and realities well beyond our comprehension and control. And when I mention science, I'm not referring only to astrophysics and its study of gravity.

"People often act in order to realize desired outcomes," observes Dutch psychologist Ruud Custers, who specializes in what science calls our *unconscious will*. "But the field now challenges the idea that there is only a conscious will. Our actions are very often initiated even though we are unaware of what we are seeking or why."[16] In plain English, Custers is claiming that we often do and say things that we cannot explain. We are in the dark about *what* or *who* is truly driving us and *why*.

In Ephesians 6:12 Paul sheds some light on this profound darkness: "For our struggle is not against flesh and blood, but against the rulers, against the authorities, against the powers of this dark world and against the spiritual forces of evil in the heavenly realms." According to both science and the Bible, in other words, the sum total of our experiences — everything we learn during our entire lives, *everything* — only hints at what's really going on. At what's really out there. At what's really true.

This is why there commonly is a striking disconnect between our dreams and our destinies, between what we set out to do and what actually happens. It's also why, when our dreams go bust, we ought not to react as if the world is coming to an end. Why, when something goes *wrong* — even catastrophically wrong — we ought to consider the possibility that something *right* is about to happen — something different, very different than we could ever imagine or pull off on our own.

Has your life turned out the way you planned, the way you dreamed it would? I'm betting the answer is no. Mine sure hasn't, and neither did my paternal grandfather's, the man for whom I'm named. He was born in 1896 in Los Indios, Texas. He had a third-grade education and no money. In his twenties, he married a woman named Dolores, who died three years after giving birth to my father.

Grandpa heard one day that there was work to be had picking crops up north. Desperate for an income, he dropped Dad off with the grandparents and hopped on a train bound for Texas City. Not long into the trip, he realized he'd forgotten his coat, so he got off the train in Kingsville. He phoned kin back home, who promised to put his coat on the next train.

While waiting, Grandpa heard music, which he followed to a storefront on Main Street. It was a church service. He wasn't a religious man, but having nothing else to do, Grandpa stepped inside and stood in the back.

That's when it happened.

A man in the congregation rose to his feet and explained in passionate language how Jesus had completely reformed his life. Grandpa recognized this man. He was from Los Indios, a very bad hombre whose apparent transformation constituted a real-life miracle. When the service ended, my grandfather hung around because he wanted to talk with the minister. If an evil homey could be transformed so dramatically, he was thinking, there had to be something to this Jesus.

Grandpa's conversation with the pastor — the entire serendipitous experience of that day — affected him so profoundly that right there he dedicated his life to Jesus. After his coat arrived, rather than continuing northward, Grandpa returned home and began preaching the Good News to migrant workers crossing the Mexican border, men and women searching not only for work but also real hope in life.

In the years that followed, one unexpected thing after another happened to my grandfather. Besides preaching, he took to collecting unusual plants, writing poetry, composing hymns, and playing a variety of musical instruments. The craziness peaked in 1937, when he was named president of the Latin American Council of Christian Churches, the oldest independent, Spanish-speaking Pentecostal organization in the nation. Today it has hundreds of churches in North America, Mexico, and Central America; and my late grandfather is still revered as a great leader, a humble servant of God, and a true renaissance man.

Who could have foreseen that Grandpa's dream to find a menial job in Texas City would lead him to discover his destiny? Only one. Isaiah reports it this way:

> "For my thoughts are not your thoughts,
> neither are your ways my ways,"
> declares the LORD.
> "As the heavens are higher than the earth,
> so are my ways higher than your ways
> and my thoughts than your thoughts."
> —Isaiah 55:8

Seeing is believing? According to the Bible, *faith* is a form of seeing, as complementary to logic as rods are to cones. That is the second lesson here for you and me.

If we live by the maxim that "seeing is believing," then we are missing out on 95 percent or more of what the universe — what God — has to offer us. *Believing* in God's infinite wisdom and love for us is our surest way of *seeing* the vast possibilities in life — not just our own home-cooked dreams.

Faith is fundamental to the Bible but also to science. Listen to the words of Saul Perlmutter, one of the astrophysicists who discovered the acceleration of the universe: "The universe is made mostly of dark matter and dark energy, and we don't know what either of them is."[17] Perlmutter has faith that dark matter and dark energy actually exist, even though the only evidence for them is circumstantial. He has faith that the scientific method provides him with all the tools he will need to understand dark energy and dark matter. He has faith that if he and his colleagues try long and hard enough, they *will* figure it all out.

That is faith far larger than a mustard seed. It is faith as large as the universe itself.

Finally, I believe an important lesson of this chapter is that the universe — you and I — are not accidents.

One of the most astonishing discoveries astrophysicists have made in recent decades is that if gravity were just 0.000000000001 (*one-trillionth* of one) percent stronger, our universe would have reversed course long ago. It would have collapsed catastrophically, ending in a *big crunch*, the opposite of the big bang. Likewise, if gravity were just 0.000000000001 (one-trillionth of one) percent weaker, our universe would have flown apart so rapidly that planets, stars, galaxies — all the basic constituents of the universe — would never have had a chance to coalesce. We'd all be dust in the wind.

Is it an accident that everything turned out so well? That gravity is not too strong, not too weak, but just right? That the big bang blew up into something so utterly propitious?

Sir Fred Hoyle, the late University of Cambridge astronomer and avowed atheist, didn't think so, not for a second. He derided the whole idea of the big bang, a term he coined on March 28, 1949, during a BBC radio program. As he put it, the big bang is an "irrational process" that "cannot be described in scientific terms."[18]

I had the privilege of getting to know Sir Fred during his years as a visiting professor at Cornell. He was famous for his strong opposition not just to the big bang, but to all theories of abiogenesis — the belief that life on earth arose by accident from nonliving chemicals.

After doing innumerable computations, Hoyle discovered that the odds of our being accidents of nature are comparable to the likelihood of a tornado sweeping through a junkyard and assembling scrap metal into a fully functioning Boeing 747. "So small as to be negligible," he said, following his calculations, "even if a tornado were to blow through enough junkyards to fill the whole universe."[19]

"One arrives at the conclusion," Hoyle declaimed, "that biomaterials with their amazing measure or order must be the outcome of intelligent design."[20] The psalmist agrees, stating the matter this way: "I am fearfully and wonderfully made; your works are wonderful, I know that full well" (Ps. 139:14).

In summary, science and the Bible agree that there is way more to life than meets the eye. Insist on living by the old adage "seeing is believing," they warn, and we remain blind to most of what exists out there, both in heaven and on earth.

Together science and the Bible enable us to see in the dark — to catch sight of truths that might sound far-fetched but are in line with what little we can see and can help guide us to the light.

Together science and the Bible inform us that the universe and you and I are special, *very special*, and that beyond our wildest imagination are worlds waiting to be explored and experienced — in this life as well as the next.

CHAPTER 5

NOT OF THIS WORLD

LIGHT IS UNEARTHLY

Light brings us the news of the Universe.
SIR WILLIAM BRAGG

God is light; in him there is no darkness at all.
1 JOHN 1:5

HAVE YOU EVER INTERACTED with something extraterrestrial?

In 1996 NASA scientists stunned the world by announcing that they had found evidence of fossilized life inside a meteorite originating from Mars. Called ALH84001 because it was found in the Alan Hills region of the Antarctic, the Martian fragment reportedly contained microscopic tube- and egg-shaped structures the scientists claimed were the ancient remains of nano-sized organisms. If true, it would prove that we aren't alone in the universe.

Excited, I flew to NASA headquarters in Houston to cover the press conference *and* to actually handle the rock — the only reporter being given permission to do so. Immediately upon arrival, I was spirited away to the restricted facility that houses the moon rocks Apollo astronauts brought back to earth.

After taking an air shower and donning a head-to-toe bunny suit, I was escorted into the capacious clean room and right up to the meteorite. The dusky, pitted, glassy, fist-sized rock was enclosed in a glove box to protect it from us and us from it.

A lively debate continues about what the meteorite's internal, fossil-like structures are, really. But I will always remember the eeriness I felt when handling an object that had traveled to our planet from so far away — an object that was truly alien.

You might never get to handle a headline-making Martian meteorite, but you do interact routinely with something that is not of this world: light. Light from the sun, light from the moon, light from stars trillions upon trillions of miles away. That's as alien as it gets.

But it is not just because of its remote origins that the aforementioned light is extremely foreign. That is the least of it. Even light coming from our bedside lamps is otherworldly. More so even than ALH84001.

Indeed, as we are about to see, science and the Bible agree that light belongs to an altogether unique category of reality. It is exceedingly exotic and operates on a privileged plane of existence that we ourselves can never experience — at least not in this life.

Science

For most of its history, science was greatly in the dark about light. Light didn't behave anything like ordinary, down-to-earth energy or matter. It was an odd sort of beast. *Very* odd.

For one thing, the speed of light appeared to be infinite. Or as the ancient Greek philosopher Epicurus put it, "quick as thought."

Quick as thought. That is not off the mark. But neither Epicurus nor any of the brilliant minds of ancient Greece fully apprehended just how much this single feature — its extraordinary speed — would set light apart as a truly alien phenomenon. That realization is commonly dated to around 1638, the year Galileo Galilei actually set about to *measure* the speed of light.

Like others of his day, Galileo believed that light traveled infinitely fast. In his epic publication *Two New Sciences*, his fictional character Simplicio says this: "Everyday experience shows that the propagation of light is instantaneous; for when we see a piece of artillery fired at great distance, the flash reaches our eyes without lapse of time; but the sound reaches the ear only after a noticeable interval."[1]

Galileo's attempt to clock the speed of light was simple and clever. He and an assistant stood on separate hilltops less than a mile apart, each holding a lantern bright enough for the other to see. Whenever Galileo uncovered his lantern, the assistant instantly did likewise with his own lantern.

Using the best clock available to him in those days — a clepsydra (water clock), most likely — Galileo recorded the time interval between when his lantern light flashed and he first saw his assistant's lantern light. After taking into account their physiological reaction times, he came up with a rough estimate of the time it took the light to travel from one hilltop to the other.

When I say "rough," I mean it. The only thing Galileo was able to say for sure was that light traveled at least ten times faster than sound. As for its actual speed, he affirmed: "If not instantaneous, it is extraordinarily rapid."[2]

Later that century, in 1676, Danish astronomer Ole Rømer became the first person ever to come up with a reliable estimate for the "extraordinarily rapid" speed of light — roughly 200,000 kilometers per second. He did it by timing how long it took light from Io, one of Jupiter's moons, to reach Earth.

Today we know with incredible accuracy that absent any air to slow it down — that is, in a vacuum — light travels at 299,792.458 kilometers per second, or 186,282.397 miles per second. That is nearly 900,000 times faster than sound! That incomprehensible speed alone sets light apart from anything else in the universe. But there is much more to the story — the surprising punch line — that science didn't realize until June 30, 1905, the day Einstein's special theory of relativity was published.

Here is the bombshell in Einstein's own, understated words (translated into English): "Light is always propagated in empty space with a definite velocity c which is independent of the state of motion of the emitting body."³ In plain language, the speed of light is unaffected by point of view — a singular privilege not enjoyed by any other known speed. The speed of light is, in a very real sense of the word, *sacred*.

Let me give you a quick illustration — in which I assume, albeit unrealistically, that everything happens in a perfect vacuum. Picture yourself standing next to a freeway where a car carrying your cousin is traveling at 60 mph. Suppose a police motorcycle whizzes past in the same direction. You and your cousin will *not* agree on how fast it's going. According to you, the police officer whipped by at, say, 70 mph. Your cousin — owing to her own 60 mph speed — will swear it was only 10 mph.

Now picture the same scenario, but with a light beam instead of the motorcycle. In this case — and *only* in cases involving light — you and your cousin will agree completely. You will both claim the light beam whizzed past at precisely 299,792.458 kilometers per second — no more, no less.

According to science, the speed of light in a vacuum is a constant of the universe — an absolute truth, not subject to opinion or any alteration. No matter who you are or what your circumstances might be — no matter what you believe or don't believe — the speed of light in a vacuum is always and forever, indisputably, immutably 299,792.458 kilometers per second, period.

And there's more. According to special relativity, the speed of light is not only sacrosanct, but also serves as an impenetrable boundary — a Great Wall of China throughout the cosmos — between two very dissimilar realms.

We ourselves inhabit the *tardyon* realm — the world of ordinary matter that can never reach or exceed the speed of light *in vacuo* from below. No matter how hard we tardyons try or how clever we are, there is no way for us to reach or break through the luminous barrier.

This is a hard pill to swallow for a species accustomed to surmounting supposedly insuperable hurdles. In 1954 Roger Bannister made history by breaking the so-called four-minute wall, running a mile in just 3:59.4 minutes. "The earth seemed almost to move with me," he recounted. "I had found a new source of power and beauty, a source I never dreamt existed."[4]

Likewise, on October 14, 1947, Chuck Yeager boarded the bullet-shaped X-1 and broke through the legendary sound barrier, unexpectedly jump-starting America's space age. "There was no buffet, no jolt, no shock," Yeager recalled later. "Above all, no brick wall to smash into. I was alive."[5]

But modern science is very clear about the light barrier; mere mortals simply aren't made of the right stuff to break through it. Here's how Einstein's equations explain it: The faster tardyons travel, the more massive they become, and therefore the more difficult it is to further increase their speed. It is a surprising quirk of nature that has been verified experimentally.

The mass of a subatomic particle whizzing around in a giant, donut-shaped accelerator — like Cornell's Wilson Synchrotron — enlarges indefinitely with increasing speed. Ultimately, an *infinite* amount of energy would be needed to propel the infinitely massive subatomic particle up to that magical, mystical, unattainable speed of 299,792.458 kilometers per second.

The second realm comprises *tachyons*, hypothetical entities that travel faster than light but cannot ever attain or break the light barrier from above. As they attempt to slow down to the hallowed speed limit, their masses inflate uncontrollably. Ultimately, an infinite amount of energy would be needed to slow one all the way down to 299,792.458 kilometers per second.

And that's not all. For tardyons and tachyons alike, time exists. It flows one way for us tardyons and the opposite way for tachyons, but both of our kinds are trapped by it. Both of our kinds are secular creatures.

Also for our two kinds, *time slows down* as we approach the speed of light — a prediction of special relativity that also has been confirmed by meticulous observations of fast-moving subatomic particles. For a tardyon traveling at 99.99999 percent the speed of light *in vacuo*, a single second stretches into a conventional day and a half. And that leads to the following conclusion: For light itself — which travels at 100 percent the speed of light — time slows to a complete stop. Time doesn't flow. Time doesn't exist. Light and light alone inhabits a realm where past, present, and future have no meaning because the three exist all together and at once.

Given all this, it strikes me as utterly crazy that although light is unattainable by us tardyons, and unfathomable as well, we interact with it daily. We rub elbows with — even internalize through our eyes and skin — the most otherworldly thing in the universe without giving it a second thought!

The craziness doesn't stop there. On September 27, 1905 — just three months following the publication of the special theory of relativity — Einstein struck again with yet another revolutionary paper, this one innocuously titled "Does the Inertia of a Body Depend upon Its Energy Content?" The gist of his proposed answer was this: Matter and energy — once thought to be as disparate as apples and oranges — are, in fact, interchangeable. Here it is in Einstein's own words: "Mass and energy are both but different manifestations of the same thing — a somewhat unfamiliar conception for the average mind."[6]

Mathematically, Einstein was saying that $E = mc^2$. But was Einstein correct? Yes, he was.

The most convincing and devastating proof of it came on July 16, 1945, at precisely 5:29:45 a.m. That is when scientists working on the top-secret Manhattan Project in Alamogordo, New Mexico, successfully detonated the world's first atom bomb. A fist-sized, 13.7-pound ball of plutonium instantaneously erupted into the energy equivalent of 40 million pounds of TNT.

With regard to our relationship with light, this interchangeability of mass and energy has two huge implications. First, it means we mere tardyons can actually become light; the only rub being that we need to die first in order for our bodies to be converted completely into pure, luminous energy. Second, it means the opposite can happen as well:

in principle, light can be transformed into ordinary matter — even flesh and blood.

These two possibilities are not merely theoretical. Nowadays, scientists can and do make them happen repeatedly and without much effort. The two processes are called "pair annihilation" and "pair creation."

In pair annihilation, an electron collides head-on with a positron, its antiparticle. Result? The two particles — the two tardyons — annihilate and become light. In pair creation, light collides with light. Result? The light disappears and rematerializes in the form of two tardyons — an electron and positron.

Taken together, the two discoveries science has made about light are truly extraordinary. The first — that light is in a realm by itself — is amazing enough. But the second — that light and ordinary matter are somehow interchangeable — is positively mind-blowing.

These discoveries suggest a mysterious yet profound connection between light and us tardyons. A connection illustrated by the miraculous twin processes of light creating matter and of matter becoming light — a connection that is devilishly hard for us to comprehend but easy for us to see with our own eyes.

The Bible

"Jesus ... said, 'I am the light of the world. Whoever follows me will never walk in darkness, but will have the light of life'" (John 8:12). Throughout human history, the identities of gods have been linked with many things — fertility, war, gluttony, you name it. But in the Bible, God and Jesus are repeatedly identified with *light*.

Scripture states the equivalence so emphatically — "God is light; in him there is no darkness at all" (1 John 1:5) — that I have long since concluded it represents not just a pretty metaphor, but an amazing truth worthy of serious contemplation. A truth remarkably in line with our modern scientific understanding of light — starting with light's privileged standing with respect to time.

"In the beginning God ..." (Gen. 1:1). The first four words of the Bible reveal that God is eternal. Even before he commences with the stupendous acts of creation memorialized in Genesis ... before he creates space and time and matter and energy ... before the appearance of earth and sky, light and dark, good and evil ... *he* exists. In the beginning God and God alone exists.

In Psalm 90:2, David rhapsodizes:

> Before the mountains were born
> or you brought forth the whole world,
> from everlasting to everlasting you are God.

There is a common misunderstanding — even among seasoned Christians — about what exactly is meant by saying that God is everlasting or eternal. The error arises naturally from our temporal view of things. It usually involves imagining God existing at every point along a timeline that stretches from negative infinity to positive infinity. This is an understandable, human-centric interpretation that we routinely impose on verses such as Revelation 1:8: "'I am the Alpha and the Omega,' says the Lord God, 'who is, and who was, and who is to come, the Almighty.'"

I believe, however, that it is more accurate to recognize that God, like light, is not a denizen of time at all. He doesn't cohabitate with us temporally — he is not imprisoned by a timeline, as we are — but is an otherworldly Presence who exists *outside* of time. After all, God existed before he created time. For the Judeo-Christian God, as for light, there is no past, present, or future. They are all one and the same to him. Past, present, and future are strictly terms we tardyons invented to help us process and talk about our time-bound reality. It is impossible for us to comprehend God's timeless ("eternal") perspective on things.

Whereas we ourselves are always developing, always becoming something other than what we were yesterday — or even a moment ago — God simply is. This explains why he identifies himself to Moses at the burning bush in a way that sounds odd to us: "I AM WHO I AM.

This is what you are to say to the Israelites: 'I AM has sent me to you'" (Ex. 3:14).

I AM. In that simple appellation, God is affirming his timelessness.

The Bible's God is equivalent to light in yet another way. The opening chapters of Genesis reveal that he is the sole author of creation and that his handiwork is fundamentally "good." Those two truths set the Bible apart from religions that teach dualism, belief systems in which adherents view the universe as having been created by dueling deities — one good, one evil. As a result, the universe is seen as being neither good nor evil, but fundamentally and irrevocably conflicted.

Dualism is a defining feature among most religions originating in the East. One such example is Zoroastrianism, the primary religion of the Persian Empire. It was founded by the prophet Zarathustra several centuries before Christ. Then, several centuries after Christ, a Persian named Mani, claiming to be an apostle of Jesus, tried hijacking Christianity by recasting it in terms of a Zoroastrian-like dualism. The heresy, widely known in those days as Manichaeism, eventually petered out, along with many other attempted bastardizations of Jesus' message.

It is not that good and evil aren't central themes in the Bible — of course they are. But the Bible speaks unequivocally about there being only one supreme deity — only one Creator — not two. As the prophet Isaiah says, God takes full responsibility for everything; the buck always stops with him: "I form the light and create darkness, I bring prosperity and create disaster; I, the LORD, do all these things" (45:7).

In the Bible, we find both God and Satan, but they are not equal and opposite deities. Like Adam and Eve, Satan is a created being who made a bad choice early on. Throughout all the drama, throughout all eternity — wrongly perceived by us tardyons as comprising a past, present, and future — God alone is sovereign.

I compare dualism's view of evil to ink. All it takes is a single drop to corrupt an entire universe of pure water. Adding more pure water will dilute the evil, but it can never remove it, never defeat it. Fully restoring the universe of pure water is impossible; evil striving against good is forever present.

The Bible compares evil not to ink but to darkness. Darkness obscures creation the way a starless night blinds us to our surroundings, but it doesn't have the power to destroy it. Moreover, darkness can be vanquished completely by light. And one day, according to biblical prophecy, the light will utterly destroy Satan and his kingdom. To wit, 2 Thessalonians 2:8: "And then the lawless one will be revealed [come to light], whom the Lord Jesus will overthrow with the breath of his mouth and destroy by the splendor of his coming."

In dualism, we are seen standing in the middle of an everlasting fight between equal and opposing forces. In the past, Hollywood filmmakers promulgated this worldview with plots that pitted bad guys in black hats against good guys in white hats. Alas, today's moviemakers love blurring the distinction between good and evil altogether.

In the Bible, our relative positioning to good and evil fits less well with Hollywood's dualism and better with special relativity's three differentiated realms of existence. In the role of light — occupying a central, privileged place in the universe — is God. On one side of him, in partial light, are we humans who, according to the psalmist, God has "made a little lower than the angels" and crowned "with glory and honor" (Ps. 8:5). On the other side of God, in total darkness, are Satan and his followers.

In this compelling picture of reality, we don't stand between good and evil, but God stands between us and evil. For us to indulge in evil, we must push past God by ignoring him, defying him, or denying him. God is the only one standing between us and total darkness. He is our one and only Savior.

There is one final way that God is identical to light. Every Christmas, the world's two billion Christians celebrate an event every bit as astonishing, as revolutionary, as the publication of Einstein's brilliant revelation concerning the interchangeability of energy and matter. Specifically, Christmas commemorates the improbable but very real interchangeability of light and tardyons. The apostle Luke describes it vividly: "The angel said to them, 'Do not be afraid. I bring you good news that will cause great joy for all the people. Today in the town of David a Savior has been born to you; he is the Messiah, the Lord'" (2:10–11). The birth of Jesus is the equivalent of pair creation — the process by which light becomes matter. In the city of David, on that extraordinary day that changed the world forever, God became human.

As for pair annihilation, one day, the Bible informs us, its equivalent could conceivably happen to us. But just as with the scientific process, certain conditions must be met. Not all subatomic particles that collide are transformed into radiant beings — and not all of us will be either. According to Christianity, when we die — when we are annihilated in this world — we will either push past the light and join evil in utter darkness or we will follow the light and join up with our eternal Creator. *It's a choice we make ahead of time in this life.*

According to the New Testament, in other words, heaven and hell are realms every bit as authentic as the tardyon and tachyon worlds described by special relativity. Two distinct afterlives. Two distinct choices. Two self-appointed fates as different as night and day.

What Does It All Mean to You and Me?

People who are on their deathbeds or have near-death experiences sometimes claim to see a bright white light just before dying. Is the phenomenon further evidence of what we have talked about here? Perhaps. One day you and I will know for sure.

For now the absolute truths about light revealed by science and the Bible convey lessons that are enormously important to us on this side of the grave. Indeed, the first, most obvious lesson is that light is necessary not only to enable our eyes to see, but for our minds, bodies, and spirits to *live*.

Science has done a fabulous job of helping elucidate the critical importance of sunshine to us and our planet. Deprived of sunshine, plants will wither and die — along with all the animals sustained by them.

Deprived of sunshine, we can become depressed as well. The affliction is called SAD — seasonal affective disorder — and psychologists treat it primarily with large doses of light.

In a paper published in the *Journal of Nutrition*, Michael F. Holick at the Boston University School of Medicine warns that a lack of sunshine is causing a serious public health problem. We are spending too much time indoors, he says, and when we do go outdoors, we cover ourselves with lotions that block out the sun. The result is that our bodies aren't producing enough vitamin D.

"The major source of vitamin D for both children and adults is from sensible sun exposure," Holick points out. But because of an epidemic vitamin D deficiency in the United States, people are at greater risk for "type I diabetes, multiple sclerosis, rheumatoid arthritis, hypertension, cardiovascular heart disease, and many common deadly cancers."[7]

Like science, the Bible has been extremely effective at elucidating the critical importance of "Son shine" to our lives. The New Testament explains that if we reject God's offer of forgiveness — delivered to us personally by his Son, his human incarnation — we are at risk of hardening our hearts, the one unpardonable sin. Like slathering on sunscreen, we can block God's light from our lives.

Consider carefully the words that John uses in John 3:19–21: "This is the verdict: Light has come into the world, but people loved darkness instead of light because their deeds were evil. Everyone who does evil hates the light, and will not come into the light for fear that their deeds will be exposed. But whoever lives by the truth comes into the light, so that it may be seen plainly that what they have done has been done in the sight of God."

A second practical application for you and me is tied up with God's famous words in Genesis 1:3: "Let there be light." I find it remarkable — at first blush, rather odd — that light should be God's initial creative act. Why didn't he begin with "Let there be space and time"? After all, those two components provide the warp and woof of the blank canvas on which God painted the universe. But no, the Creator assigned top priority to light. Why?

It is because "Let there be light" makes the declaration that God will be an integral and indelible part of the cosmos. "Let there be light" lets us know from the get-go that God is painting himself into the picture. With the command "Let there be light," God is proclaiming his sovereignty over creation. Declaring that he is no arm's-length deist-God, but a relational God who will take an active interest in the affairs of our cosmos. Assuring that no one who scrutinizes the universe — or even looks upon it casually — can fail to see him in it.

The Nobel Prize–winning physicist Sir William Bragg said, "Light brings us the news of the Universe."[8] And the Bible agrees. Light does bring us the news — the *good* news — of the universe, directly from the Creator himself. The message for us is crystal clear. Anyone who claims they can't see God and therefore can't believe in him is being willfully blind. Romans 1:20 says it beautifully: "Since the creation of the world God's invisible qualities — his eternal power and divine nature — have been clearly seen, being understood from what has been made, so that people are without excuse."

The third practical application concerns theodicy, our imperfect and frustrating attempt to understand how evil managed to creep into God's thoroughly "good" creation. It is a matter of great importance to us, for we must cope with evil every day. I support the hypothesis that evil becomes possible as soon as one allows free will into the picture — a truth that is entirely consistent with both science and logic. But the main theodicean lesson we derive from our discussion here is this: God is not waiting around to clean up the deadly mess; he has already eliminated evil from the universe. We just haven't witnessed it yet. Remember, it is *we* who are stuck in the timeline, not God.

Suppose we are driving north on the expressway and see an accident take place suddenly on the southbound side. Predictably, vehicles using the southbound lanes hit the brakes, and traffic on that side begins to back up — perhaps even for miles. As we continue driving northbound unimpeded, we have a perspective on things that the southbound drivers do not. We have seen their future. We are in a position to tell them that the backup is temporary and that beyond the accident there is smooth sailing.

That is precisely what our eternal Father knows about our own situation and the truth that Jesus communicates to us throughout the New Testament. In John 16:33 he tells us flat out, "I have told you these things, so that in me you may have peace. In this world you will have trouble. But take heart! I have overcome the world."

The final lesson for you and me, I believe, is this: science and the Bible both exist to fight against darkness and promulgate enlightenment. Both of them urge us not only to *see* the light but to *be* the light as well. "As far as we can discern," Carl Jung once mused, "the sole purpose of human existence is to kindle a light in the darkness of mere being."[9] Likewise, the apostle Paul exhorts us in Ephesians 5:8: "You were once darkness, but now you are light in the Lord. Live as children of light."

CHAPTER 6

AN EGG-STRAORDINARY EVENT

THE UNIVERSE WAS CREATED *EX NIHILO*

*Through him all things were made; without
him nothing was made that has been made.*

JOHN 1:3

In the beginning there was an explosion.

STEVEN WEINBERG

"OUT OF THIN AIR..."

To an amateur magician like me, that familiar phrase carries with it many memories of astonishing tricks I have performed over the years. My specialties are escape and close-up magic, and one of my favorite illusions is making an egg materialize from my seemingly empty mouth. It never fails to elicit lively applause.

Ordinary people have magic-like experiences every day, too. My grandfather Miguel Guillén, after whom I'm named, comes to mind. Despite being an elementary school dropout; despite becoming a widower and single dad early in his marriage; despite living during the Great Depression — against all odds, he achieved a greatness no one could have foreseen. He became one of the founders of the Spanish-speaking Pentecostal movement in the United States and Mexico, wrote poetry, composed hymns, and mastered the violin.

Grandpa isn't alone. I know of many men and women who, like the legendary Horatio Alger, went from rags to riches. Individuals who, like magicians, appear to have created something out of nothing.

Billy Graham, the son of North Carolina dairy farmers, created a Christian ministry that continues serving and inspiring people worldwide. "Graham has been credited with preaching to more individuals than anyone else in history," the editors of *biography.com* state, "not counting the additional millions he has addressed through radio, television and the written word."[1]

In this chapter we will look behind the scenes of what is easily the most magical rags-to-riches story of all time: the origin of the universe. We will attempt to answer the question "Where did everything that we see (and don't see) come from?" And we will discover that until the early twentieth century, science was extremely confident it knew the answer, but turned out to be dead wrong and needed to start over.

Today both science and the Bible agree on three amazing truths concerning this important question. First, they agree that the universe had a definite beginning. Second, that it materialized — not even out of thin air — but out of *absolutely nothing*. And third, that "absolutely nothing" is an illusion; it is not actually nothing. In truth, it is something more incomprehensible even than the universe itself.

Science

In the beginning ... there was *no* beginning. Heading into the twentieth century, that was science's strongly held position concerning the origin of the universe.

According to the then-popular "steady state" hypothesis, the cosmos always was and always will be exactly as it exists right now. It was an idea that, among other things, enabled science to distance itself from Genesis, from any discussion of a creation or creator. For if the universe is indeed eternal and static — if it shows no evidence of expanding or contracting — then there is no seminal moment when it sprang into existence.

Back then the stature, credibility, and reliability of the steady state hypothesis seemed above questioning — as evolutionary biology seems to be today. Anyone who dared to disagree with it could expect to reap enormous resistance and even ridicule. Just ask Georges Henri Joseph Édouard Lemaître. He was a brilliant Belgian cosmologist with *two* PhDs — one in mathematics from the Catholic University of Louvain and one in physics from MIT. He was also a Roman Catholic monsignor. In 1927 he dared to question the steady state hypothesis and was dressed down for it by none other than Albert Einstein.

Here is what happened. In 1915 Einstein created a scientific commotion by publishing the general theory of relativity, his groundbreaking take on gravity. (For more on this, see chapter 4.) Thus, many of science's cherished notions about the universe went out the window. But not the steady state hypothesis. Against all odds, it held steady (forgive my pun) against the onslaught of Einstein's epiphanies.

But not for long. In the 1920s two scientists discovered solutions to Einstein's gravity equation that allowed for the shocking possibility of an *expanding* universe. The revelation came separately from Russian physicist Alexander Friedmann and Lemaître. They published their results in different years (1922 and 1927, respectively), different countries, and different languages — each one unaware of the other.

Being a mathematician not much concerned with physics or the real world, Friedmann didn't make a big deal of the expanding-universe solution. And frankly neither did anyone else — including Einstein, who pooh-poohed it as a hypothesis not to be taken seriously.

Think of it this way: It is mathematically possible that one day you will discover that every single item on your grocery list has been marked down by 30 percent. But how likely is that to happen? Not very. Such was the reaction most scientists had toward the expanding universe solution of Einstein's equations: interesting but not realistic.

Lemaître's attitude, however, was quite different. Excited by the possibility that science might be mistaken about the universe not having a beginning — excited by provocative new evidence that galaxies (nebulae, they were called back then) were moving away from one another — the cosmologist-monsignor wrote a paper in 1927 titled "A Homogeneous Universe of Constant Mass and Growing Radius Accounting for the Radial Velocity of Extragalactic Nebulae." It was published in French and appeared in a little-known Belgian astronomical journal, so not many scientists were influenced by it. But wherever he went, Lemaître never failed to lobby for the idea that the universe was not static after all, but expanding.

One of those occasions was in October 1927, when the world's most brilliant scientists gathered in Brussels for the Fifth Solvay International Conference. Its illustrious attendees included no fewer than seventeen physicists and chemists who had won or would win the Nobel Prize. Here, in Lemaître's backyard, was a golden opportunity for him to promulgate his pet theory.

As a scientist, I have attended similar gatherings and know firsthand how intimidating they can be. In the midst of so much brainpower, egoism, power-mongering, the entire smorgasbord of human frailties, it is difficult to get a word in edgewise, especially about an unpopular idea. But somehow the persistent Lemaître managed to do it, and the response he received from Einstein, hands down the superstar of the conference, was blistering. "Your calculations are correct," Einstein is reported to have sneered, "but your grasp of physics is abominable."[2]

Ouch.

Still, Lemaître refused to blink. In fact, he felt emboldened in 1929 by the bombshell publication of the American astronomer Edwin Hubble, which left no doubt that galaxies were moving away from one another. Rightfully so, Lemaître considered it to be a stunning validation of his uninfluential paper of two years earlier.

In 1931 he upped his assault on scientific orthodoxy during a meeting of the British Science Association, convened to discuss the spiritual implications of science's study of the universe. Lemaître argued that if the universe were indeed expanding, then there must have been a creation moment in its deep past when everything was set into motion. He posited an instant when it all sprang forth out of some kind of seed — in his words, a "primeval atom."[3]

Later that year, Lemaître published a further explanation of his origin theory in the prestigious British journal *Nature*. In it he made use of a variety of metaphors — at one point likening the expanding contents of the universe to "cosmic fireworks" and the primeval atom to a "cosmic egg exploding at the moment of creation."[4]

Here is how he envisioned the sequence of events: "This [primeval] atom is conceived as having existed for an instant only.... As soon as it came into being, it was broken into pieces which were again broken, in their turn; among these pieces electrons, protons, alpha particles, etc., rushed out.... The disintegration of the atom was thus accompanied by a rapid increase in the radius of space which the fragments of the primeval atom filled."[5]

Throughout the 1930s and into the 1940s, Hubble's evidence of expanding galaxies gradually persuaded scientists to soften their position against an expanding universe — but did nothing to win them over to the idea of a definite origin. "Philosophically," the eminent astronomer Sir Arthur Eddington opined, "the notion of a beginning of the present order of Nature is repugnant to me.... I should like to find a genuine loophole."[6] The dilemma for Eddington and science was daunting, to say the least. How could they continue believing in an expanding universe but not in a creation moment? Still, never underestimate the ability of the human mind to rationalize its way out of any mess.

In 1948 three scientists — Thomas Gold, Hermann Bondi, and Fred Hoyle — published a novel theory of the universe that was part steady state and part expanding. It imagined a universe that had always existed and would always exist — but that was constantly stretching. As it enlarged, new material appeared out of nowhere to maintain a constant, uniform density. Yes, the authors conceded, their theory violated the law of conservation of energy and matter, but by an amount too small to be detected experimentally.

By then, a smattering of scientists — most notably, the eminent Russian cosmologist George Gamow — had taken up Lemaître's cause and further refined the cosmic egg theory. But during a BBC radio broadcast in 1949, while defending his own expanding steady-state scenario, Hoyle disparaged Lemaître's and Gamow's idea. He referred to it as "the hypothesis that all the matter in the universe was created in one big bang at a particular time in the remote past."[7] Hoyle's cynical phrase "big bang" stuck.

From then on, the two cosmological theories competed for science's favor without any clear winner. It wasn't until 1964, when a pair of American scientists discovered an aura of microwave radiation throughout the heavens — the afterglow, they argued, of the universe's explosive past — that most scientists finally capitulated and embraced the big bang theory.

Lemaître didn't hear about the discovery until 1966, by which time he was in his late seventies. Two weeks later he passed away.

Scientists since then have continually monkeyed with the big bang theory — in certain ways, considerably so — and even rebranded it. Nowadays, the canonical description of the origin of the universe involves the preferred and more accurate name of *inflation theory*.

Nevertheless, Lemaître's original, basic notion has survived the extensive modifications and, because of it, scientists still struggle mightily to answer two questions: Where did all the stuff that went into making the universe come from? And what exactly triggered the creation process?

The proposed answers are quite ingenious and rely mostly on a decidedly counterintuitive definition of "nothing." According to one favored scenario, the universe abruptly grew out of the so-called quantum vacuum, which by definition is what is left over after all matter and energy have been removed from space. How can that be? How can an entire cosmos spring from something that is void of all matter and energy?

The theory only makes sense when we realize that according to modern science, the quantum vacuum is actually far from being nothing. Rather, it is filled with all sorts of exotica — quantum fields and virtual particles — that can't be detected by any ordinary means. This makes the quantum vacuum the mother of all piñatas, the mother of all wombs, and, so says modern science, the mother of our universe.

As to what provoked the quantum vacuum to cough up an entire universe, some proposals are clearly meant to studiously avoid any idea of a creator. To wit, astronomer Stephen Hawking, a professed atheist, claims, "Because there is a law such as gravity, the universe can and will create itself from nothing."[8] The idea is wildly speculative, to say the least. But even if it turns out to be true, Hawking's proposal begs the question: How did such a truly extraordinary law of gravity come about? What created *it*?

Reflecting back on science's checkered efforts to explain the origin of the universe, I and many others have wondered what motivated Lemaître to advocate so passionately the idea of an expanding universe *with an origin*. Was it merely a gut feeling, or was it because of his religious worldview?

Lemaître himself denied the last possibility quite strenuously, claiming that he always had taken great care to keep a firewall between his religious beliefs and the scientific demand for objectivity. When Pope Pius XII declared that the cosmic egg theory validated the Catholic faith, Lemaître demurred diplomatically with these words: "As far as I can see, such a theory remains entirely outside any metaphysical or religious question. It leaves the materialist free to deny any transcendental Being."[9]

Also I wonder: *What can we surmise from science's initial, strongly negative reaction to Lemaître's proposal? Was it simply the human tendency to protect the status quo? Could it have been a product of human pride and the rampant egoism among leading scientists, which we noted earlier? Or was it evidence of a collective prejudice against the biblical worldview?*

Brian Greene, theoretical physicist at Columbia University and bestselling author of *The Elegant Universe*, has this to say: "I think there was some resistance to having somebody in the scientific camp and the religious camp. But did that make it difficult for some scientists to perhaps embrace him as much as they would someone who was not in that position? Probably."[10]

What we know for certain is this: Last century, in the space of five exciting decades, science did a historic about-face concerning its explanation for the origin of the universe. Because of the evidence, scientists now cling to the theory that the cosmos had a beginning. And that something — some eternal process or unknown physical principle, older certainly than the current, visible world — is responsible for pulling the trigger.

The Bible

Before science experienced its dramatic change of mind, atheists relied on the steady state hypothesis to bash any belief in a creator. Nowadays many of them persist in using science to proselytize, but the big bang theory makes it harder for them to make their case. It requires them and everyone else to answer the questions: Who or what set off the creation moment? Which creator should we believe in?

Most of our earliest recorded creation stories invoke a dualist, polytheistic view of reality. They portray us, our planet, and the universe as ensuing from some melodramatic competition or copulation between equally resourceful good and evil deities. (For more on this, see chapter 5.) Case in point: the Mesopotamian/Babylonian creation story *Enuma Elish*. Probably recorded around the twelfth century BC, it features a cast of named characters larger than that of a modern-day soap opera. Starring in the Babylonian version of the story are Tiamat, the evil goddess of chaos, and Marduk, the good god of order. Other versions replace the hero Marduk with the gods Anu, Enil, or Ninurta.

The opening portion of *Enuma Elish* recounts Tiamat's malevolent scheming and terrifying reign over a riotously misbehaving pantheon of gods. Right when it seems that no one can defeat her, Marduk steps up and does her in with an arrow through the heart. At this point, Marduk, the lord, creates the heavens and the earth from the two halves of Tiamat's dead body. Here is the exact account of it:

> The Lord rested, examining her dead body,
> To divide the abortion and to create ingenious things therewith.
> He split her open like a mussel into two parts;
> Half of her he set in place and formed the sky therewith as a roof....
> And a great structure, its counterpart, he established [with the other half], namely Esharra [Earth].[11]

Then comes the time for the gods to create people. They slay Kingu — Tiamat's evil husband — and use his blood to fashion us. This explains our corrupt nature and our lowly purpose in life, which is taking on the menial jobs formerly performed by enslaved, rebel gods. Here is how the *Enuma Elish* puts it:

> Kingu it was who created the strife.
> And caused Tiamat to revolt and prepare for battle.
> They bound him and held him before Ea;
> Punishment they inflicted upon him by cutting the arteries of his blood.
> With his blood they created mankind.
> He [Ea] imposed the services of the gods upon them [mankind] and set the gods free.[12]

Not even science can prove what really happened at the start of the universe. In order to be certain, we would need to witness it firsthand. The best any of us can do now is consider with an open, critical mind the evidence available from all quarters.

I have dedicated my life to doing just that, and here is what I wish to say to you. After decades as a scientist, after reading and analyzing scores of creation stories, I'm struck by the similarity of the narratives offered by science and the Bible. Unlike the fantastical, convoluted narrative of the *Enuma Elish* and many others like it, the creation stories presented by science and the Bible are straightforward and matter-of-fact. They report what happened with minimal color commentary. Additionally, both stories star a "holy trinity." For the Bible, it is the Father, Son, and Holy Spirit. For science, it is the Law of Gravity, Randomness, and Natural Selection.

Moreover, both stories explain creation as a series of significant milestones, each building logically on the one before it. And the two sets of milestones — one scientific, one biblical — are reminiscent of one another. For example, here is how science sees things getting started:

> In the beginning Randomness and the Law of Gravity created the heavens and the earth. Now the earth was formless and empty, darkness was over the surface of the deep, and Randomness and the Law of Gravity were hovering over the quantum vacuum. And Randomness and the Law of Gravity said, "Let there be a perturbation!" and there was a perturbation, followed immediately by inflation. And there was a start to it, and a finish to it — the first milestone.

Both science and the Bible elaborate in a series of subsequent milestones the increasingly complex cosmos. And they both do it in a no-nonsense, rational way — to the degree that any narrative claiming to explain the origin of a universe can sound entirely rational.

After the manifestation of atoms and stars and galaxies, and following the formation of our solar system, come the climactic milestones. Science describes them this way:

> And Randomness and Natural Selection said, "Let the water teem with living creatures, and let birds fly above the earth across the vault of the sky." So Randomness and Natural Selection — guided by the many laws of science — initiated the creation of the great creatures of the sea and every living thing with which the water teems and that moves about in it. And there was a start to it, and a finish to it — the umpteenth milestone.
>
> And Randomness and Natural Selection said, "Let the land produce living creatures: the livestock, the creatures that move

along the ground, and the wild animals." And it was so. Randomness and Natural Selection — guided by the many laws of science — made the wild animals, the livestock, and all the creatures that move along the ground. And Randomness and Natural Selection saw that it was good.

Then Randomness and Natural Selection said, "Let us make humanity in our image, in our likeness, so that they may rule over the fish in the sea and the birds in the sky, over the livestock and all the wild animals, and over all the creatures that move along the ground."

So Randomness and Natural Selection — guided by the many laws of science — created humanity in their own image, in the image of Randomness and Natural Selection they created them; male and female they created them.

Randomness together with Natural Selection blessed them and said to them, "Be fruitful and increase in number; fill the earth and subdue it. Rule over the fish in the sea and the birds in the sky and over every living creature that moves on the ground."

Randomness together with Natural Selection saw all that they had made, and it was very good. And there was a start to it, and a finish to it — the umpteenth-and-one milestone.

I have deliberately chosen to word science's creation story in a way that underscores its similarities to the Bible's creation story. And it works. That's amazing. Sure, science and the Bible appear to disagree on the order of certain milestones, the role of evolution, and the ages of everything — ostensible differences that hog the public's attention and certainly are not insignificant.

Also, whereas the Bible's account of creation is uplifting because it describes a way for us to be reconciled and reunited with our Creator, science's account is rather depressing. "The more the universe seems comprehensible," remarks physicist and atheist Steven Weinberg, "the more it also seems pointless."[13]

Science's creation story makes it clear that we are made in the image of randomness and natural selection; that is, we are the offspring of a stupendous series of accidents. The same is said to be true of the universe. According to the scientific narrative, everything we are and everything we observe around us happened by chance over a very long period of time.

This, too, is very different from the account we find in the Bible. It is the difference between a person looking into a mirror and seeing a creature who is one step removed from a chimp and one looking and seeing a creature who is one step removed from God. It is the difference between believing that accidental, terrestrial-born senses and accidental, terrestrial-born brains are somehow capable of comprehending the universe at large and believing that our senses and brains were intentionally made to comprehend not just the universe but its Creator as well.

Is the Bible's creation story merely a vanity? A way of puffing ourselves up to be more like deities and less like chimps? A way of making ourselves, our planet, our solar system, and our universe seem more special than they are actually?

As I said, not even science can prove what really happened during creation; so, yes, it is possible. Yes, it is conceivable that astronomer and agnostic Carl Sagan was correct when he said this: "For as long as there have been humans we have searched for our place in the cosmos. Where are we? Who are we? We find that we live on an insignificant planet of a hum-drum star lost in a galaxy tucked away in some forgotten corner of a universe in which there are far more galaxies than people."[14] But do I believe he is correct? Do I agree that we are an insignificant species living on an insignificant planet in an insignificant galaxy? No, I do not — and neither does science or the Bible.

According to them both, we are exceptional. Their only disagreement is whether we are *accidentally* exceptional or *intentionally* exceptional. (I will have a great deal more to say about this in chapter 10.)

I do not believe the Bible's creation story puffs us up. We aren't the heroes of Genesis, not by any stretch of the imagination. We were made in God's image, yes, but we also quickly blew it by refusing to obey a simple request of his — and from then onward dug ourselves into a deeper and deeper hole. If we really wanted to make ourselves look good, we'd destroy the Bible, not promote it.

With so much agreement between the creation stories of science and the Bible, it is only natural to wonder: *Will a day come when there is complete agreement between them?*

I believe the answer is no because, for one thing, science makes it a point not to acknowledge God. Too many people assume this means that science opposes or disapproves of God — a misimpression happily encouraged and even promulgated by many atheists. But it isn't true.

Science is agnostic on the question of God's existence. That is what Lemaître was saying to Pope Pius XII. Science is devoted to finding wholly rational (or at least, rational-sounding) explanations for everything, including the moment of creation. Invoking God, therefore, is simply not an option.

In this regard, science is like the popular board game Taboo. According to the rules, contestants need to communicate a secret word to their teammates by giving them clues. But they must never use certain forbidden words while doing so. Likewise, according to the rules of science, contestants need to explain the universe by doing carefully controlled experiments and publishing their analyses in peer-reviewed journals. But they must never use certain forbidden words while doing so — above all, the word *God*.

Thus, even on the one supreme point where the creation stories of science and the Bible appear to disagree — the issue of God's existence — they don't actually disagree. As Weinberg, the atheist-physicist, concedes, "Science does not make it impossible to believe in God. It just makes it possible to not believe in God."[15]

What Does It All Mean to You and Me?

You and I are the most creative creatures on the planet because we are made in the image of a creative God. I believe that is the first message we should glean from this discussion — and why it should both uplift and humble us.

AN EGG-STRAORDINARY EVENT

When I was a kid, my family had little money. But we never went hungry because my mother was a kitchen magician. She could whip up an inexpensive meal from scratch, and it was always tasty.

From scratch. Interesting phrase, isn't it? And misleading. It is like referring to someone as *self-made*. We invoke those phrases, of course, to describe authentic achievement. To describe how we are seemingly able to create something from nothing. Yet even the lowliest ingredients that go into an exceptional creation originate with the ultimate Creator and are far from being nothing.

Flour, sugar, and spices, when studied up close, are revealed to be magnificent, complex chemicals comprising many atoms. The essence of simple nutmeg — isoeugenol — has the chemical formula $C_{32}H_{56}$ and goes by the scientific name: *3,3-dimethyl–2-methylidenebicyclo[2.2.1] heptane;(4R,6R)–1,6-dimethyl–4-propan–2-ylcyclohexene;(2R,4R)–2-methyl–1-methylidene–4-propan–2-ylcyclohexane.*

Similarly, the meanest-seeming roots of a person's life never amount to nothing. John Marks Templeton, born in rural Tennessee to a family of modest resources, is credited with becoming a self-made billionaire — "arguably the greatest global stock picker of the century," declared *Money* magazine in 1999.[16] Yet the essential ingredients of his success can be traced back to his parents, who taught him a work ethic, an entrepreneurial spirit, and solid Christian values.

The notion of anything being truly self-made or created from scratch — from nothing, from out of thin air — is an illusion at best and a lie at worst. Science and the Bible now agree that even the universe was created from something, from some preexisting reality, although it gave the appearance of being a void.

This lesson should keep us humble, discouraging us from becoming too full of ourselves whenever we create something wonderful or we prosper in life. Whatever noteworthy thing we have accomplished, each of us is indebted to — and should practice gratitude for — the people, the circumstances (good and bad), and the God-given gifts that helped us get there.

I had the privilege of preaching at the Crystal Cathedral during the glory days of Dr. Robert H. Schuller's world-famous ministry. On one occasion, while I was sitting with him on the dais, Dr. Schuller leaned over and said, "I'm like the frog sitting on top of a very tall flagpole." Then he said, gesturing toward the congregation, "And it's because of them and God that I got there."

The second lesson from this discussion is equally important: creativity is not a random process. Taking disparate ingredients and fashioning them into anything extraordinary — be it the *Mona Lisa* or *The Brothers Karamazov* or *It's a Wonderful Life* — requires an organizing principle. It takes an intelligent or intelligent-like maestro to construct the extraordinary thing in a stepwise fashion, in accordance with some clear-cut vision or natural law.

On August 21, 2011, an American software engineer named Jesse Anderson created the Million Monkey project, which featured millions of virtual simians typing away randomly. In just forty-six days, Anderson claims, the mindless authors recreated all of Shakespeare's thirty-eight major works. "This is the largest work ever randomly reproduced," he crowed.[17]

The media trumpeted the achievement uncritically. Yet the claim is so misleading as to border on deception. In truth, what the digital monkeys produced randomly were unbroken strings of letters. It took a computer program (a digital maestro) working behind the scenes to recognize correct sequences and break them up — intelligently, not randomly — into the proper words.

My purpose is not to criticize Mr. Anderson or accuse him of any wrongdoing, nor even to lament the ignorance of the popular press. It is to illustrate our second lesson — that creativity is not a random process. Science and the Bible agree that in order to create something from nothing, there needs to be something or someone behind the scenes directing the show.

The final practical lesson for us, I believe, is this: when we remove the sights and sounds of life, the result is a silence filled with God's presence — the spiritual equivalent of the quantum vacuum. When we pay close attention to that seeming nothingness — by isolating ourselves from our noisy, everyday world — we are able to recognize great truths. Psalm 46:10 puts it this way: "Be still, and know that I am God."

I recall being on assignment in Greenland for *Good Morning America*. During my visit, a scientist invited me to jump aboard a snowmobile and follow him to a place far from our ice station, way out in the middle of nowhere. After arriving, we turned off the engines and listened. We were in a place that was thoroughly silent and white — the closest thing to nothingness I have ever experienced. According to science and the Bible, in the split second just before creation, there was a similar seeming nothingness pregnant with possibilities.

By removing ourselves from the distractions of modern life, we can replicate that remarkable seminal moment. By immersing ourselves in nothingness, we can discern its great, hidden secrets.

Get away from the bright lights of civilization and travel two miles up the slope of Mauna Loa in Hawaii. Up there at night, using a special telescope to peer into the seeming nothingness of deep space, one is able to detect the microwave aura, the first light ever created in the universe.

Escape from the loud noises of city life and head to a place of quiet, unspoiled beauty — like the Grand Tetons in northwest Wyoming or the inside of a prayer closet in your own home. There, in the solitude of that sacred space, you will be able to hear the voice of God, the Creator of the universe.

CHAPTER 7

THE CERTAINTY OF UNCERTAINTY

TRUTH IS BIGGER THAN PROOF

*Logic is an organized way of going
wrong with confidence.*

CHARLES F. KETTERING

*Guide me in your truth and teach me,
for you are God my Savior,
and my hope is in you all day long.*

PSALM 25:2

OH YEAH? *PROVE IT!*

How often we hear that demand from people seeking irrefutable evidence for something they find hard or impossible to believe. Yet, as we are about to see, science and the Bible both agree that proof — even of objective truths — is not only hard for us to obtain, but in many cases flat-out impossible.

I was at ABC News in 1994–95 when O.J. Simpson was accused of murdering his wife, Nicole Brown, and her friend Ronald Goldman. For twelve long months my network — like seemingly every news operation in the world — gave the sensational case wall-to-wall coverage. Again and again, the public was reminded of the legal definition of proof in a criminal trial. O.J. Simpson needed to be found innocent unless the prosecution successfully convinced a hand-picked jury of ten women and two men — nine blacks, two whites, and one Hispanic — that he was guilty *beyond a reasonable doubt*.

On October 3, 1995, when the jury found Simpson "not guilty," there was unbridled celebration among black Americans and stunned disbelief among white Americans. According to a CNN/*Time* magazine poll taken immediately after the trial, a mere 14 percent of blacks believed Simpson to be guilty (66 percent, not guilty), whereas 62 percent of whites believed him to be guilty (only 21 percent, not guilty) — proof that "certainty" in a legal sense is anything but certain.[1]

Further evidence of this came many years later, on the trial's twentieth anniversary. On June 9, 2014, a CNN/ORC International poll revealed that a majority even of black Americans — 53 percent — now believe that O.J. Simpson is "definitely" or "probably" guilty of the crime.[2]

Coincidentally, while Simpson was being tried for murder, I was covering a very different trial — in its own way, no less sensational — being prosecuted within the courtroom of mathematics. Andrew Wiles, a British number theorist, claimed to have proven a theorem that had resisted being solved for centuries. It is called Fermat's Last Theorem, after the renowned seventeenth-century French mathematician Pierre de Fermat. He claimed to have solved the vexing riddle, but to everyone's everlasting frustration, never actually delivered the goods.

In late 1994, when Wiles announced he'd done it, the verdict from his peers was decidedly mixed. Many didn't believe him — and those who did believe him couldn't easily defend their position because Wiles's supposed solution was extremely long and complicated. Worse for him and his assertion, in the previous year someone had discovered that Wiles had made a mistake in reasoning, a slip-up Wiles later claimed to have fixed.

It wasn't until May 1995 that the debate was finally settled. Wiles's 109-page-long proof of Fermat's Last Theorem was fully vetted by a jury of distinguished mathematicians. The panel declared Wiles's work to be free of any errors, and his paper was published in the *Annals of Mathematics*.

But that isn't the end of the story — more like the beginning of it. Today mathematical "proofs" are so long and complex — routinely comprising hundreds of densely written pages and requiring years of fact-checking by expert referees — that there is a genuine crisis. Stanford University mathematician Keith Devlin describes it this way: "I think that we're now inescapably in an age where the large statements of mathematics are so complex that we may never know for sure whether they're true or false."[3] This is shocking news for a discipline that historically has given proof a stellar name, that always has been venerated as the gold standard for judging what is objectively true or not.

Those glory days are over. Now, even mathematics has descended to the world of uncertainty all too familiar to the rest of us. As Devlin observes, "It makes it [mathematics] more human."[4]

Equally shocking news is that science and the Bible agree that our predicament is unavoidable. We live in an uncertain world where honest-to-goodness proof exists in not many places other than our imagination — and certainly not in everyday life. Ours is an uncertain world where, apart from proving some basic mathematical theorems, logic fails miserably. In this uncertain world, achieving any semblance of real clarity about the universe or ourselves requires what we call *intuition* — as well as its peculiar, widely misunderstood companion, which is far more powerful than logic: *faith*.

Science

An informal caste system exists among the various scientific disciplines. At the top of the heap is mathematics — commonly referred to as "queen of the sciences." Ironically, even though pure mathematics is not concerned with the real world, its results frequently have very practical applications, science being the greatest beneficiary. "The miracle of the appropriateness of the language of mathematics for the formulation of the laws of physics," Hungarian physicist Eugene

THE CERTAINTY OF UNCERTAINTY

Wigner once observed, "is a wonderful gift which we neither understand nor deserve."[5]

Beneath mathematics in the academic pecking order are physics, chemistry, medicine, geology, and biology — the so-called hard sciences. At the bottom are psychology, sociology, anthropology, and the like — the soft sciences.

Many considerations go into making these class distinctions, but the most important one has to do with validation — with how closely a particular science is able to achieve real proof. Mathematics sits at the very top because it alone operates in the abstract world of pure logic and therefore can prove things conclusively.

Up to a point, that is.

As a science-minded kid, I was fascinated by logic, by its seeming potency in the face of life's wishy-washiness. With logic's clear-cut, uncompromising rules, I could articulate an unassailable, air-tight defense of a truth and say to any Doubting Thomas, "Take that!" By the time I matriculated to high school, I'd accumulated a sizable library of books on the subject of logic and proof, many of them written by the incomparable polymath Martin Gardner. Among my favorite titles of his were *My Best Mathematical and Logical Puzzles*, *Perplexing Puzzles and Tantalizing Teasers*, and *Mental Magic: Surefire Tricks to Amaze Your Friends*. These books — chockablock full of logical riddles and proofs — were the start of my lifelong, loving relationship with deductive reasoning, the powerful methodology whose origins we credit to Aristotle.

The quintessence of Aristotelian reasoning — the epitome of *proof* — is the rock-solid syllogism illustrated here:

> All dogs have four legs
> Fido is a dog
> Therefore, Fido has four legs

Euclid was the first to utilize the rules of Aristotelian logic in a spectacular way. Starting with only ten assumptions — for example, if two things are equal to the same thing then they themselves are equal — Euclid deduced the whole of plane geometry. He proved for all time the scores of theorems students in high schools everywhere learn today.

At the end of the nineteenth century, German logician Gottlob Frege set about to logically deduce the whole of arithmetic in the same way that Euclid had done with geometry. It took him ten years, but he did it.

Or so it seemed.

In 1902, just as Frege was preparing to publish the last part of his great achievement — his *Grundgesetze der Arithmetik* (Fundamental Laws of Arithmetic) — mathematician Bertrand Russell spotted a problem. Not a mistake by Frege, but a basic defect in logic itself — serious enough to spell the end of not only Frege's ambitious enterprise, but eventually our entire centuries-long honeymoon with logic as well.

Russell's devastating insight, which concerned the logical study of classes of objects, would take me several pages to explain.[6] Instead, I can easily illustrate the point by asking you to consider this seemingly simple declaration:

"This statement is not true."

Do you see the problem? If the assertion is true, then it is not true. If it is not true, then it is true. It is a paradox, one of logic's many shortcomings.

"A scientist can hardly meet with anything more undesirable than to have the foundation give way just as the work is finished," Frege lamented. "In this position I was put by a letter from Mr. Bertrand Russell as the work was nearly through the press."[7]

Logic's suddenly diminished stature was further and permanently sealed three decades later by the landmark discovery of a twenty-five-year-old named Kurt Gödel, arguably the most perspicacious logician who ever lived, after Aristotle himself. In 1931 the young Austrian proved that when it comes to systems at least as complicated as arithmetic, logic crashes — like an overloaded supercomputer — and proof

THE CERTAINTY OF UNCERTAINTY

isn't always possible. Stated in plain language, Gödel's so-called incompleteness theorem means this: There will always be objective truths that we can't prove using logic. Ever.

In 1956, after a long and illustrious career, Russell reflected back on the irrevocable damage Gödel and he himself had inflicted on the once vaunted reputation of logic and proof. "I wanted certainty in the kind of way in which people want religious faith. I thought that certainty is more likely to be found in mathematics than elsewhere." But, he concluded, "after some twenty years of arduous toil, I came to the conclusion that there was nothing more that I could do in the way of making mathematical knowledge indubitable."[8]

Truth is bigger than proof.

In one way or another, this same loss of certainty has occurred in the other two classes of science as well — the hard and soft sciences. The only difference is they didn't have as far to fall. That is because from the get-go, the formulators and followers of the scientific method knew they were in the business not of proving but of demonstrating. They realized that the verdicts of science — what we call scientific theories — are supported by the evidence, yes, but also by all the personal biases of which any jury is susceptible. They realized that like the legal system, science doesn't always get it right.

Even popular, seemingly indisputable scientific theories are just one unexpected discovery away from being shot down. Einstein summarized the inevitable uncertainty this way: "No amount of experimentation can ever prove me right; a single experiment can prove me wrong."

In the hard sciences, uncertainty reared its head in 1927 when German physicist Werner Heisenberg discovered a fundamental fuzziness in the machinations of the universe. He identified a stubborn, granular vagueness that becomes more and more apparent the closer we look at such things as space, time, mass, and energy. Heisenberg's uncertainty principle is the hard sciences' counterpart to Gödel's incompleteness theorem.

Here's one way of seeing it. Suppose we want to know the pressure inside a bicycle tire. To find out, we use a pressure gauge. But the gauge works by sampling the air inside the tire, which alters the pressure ever so slightly. It therefore unavoidably introduces an uncertainty into our measurement.

According to the spirit of Heisenberg's uncertainty principle, we are damned no matter what we do. If we take a good-sized sample of air in order to get an accurate measurement, we disturb the system greatly. If we take a minuscule sample in order to minimize our meddling, we get a substantially unreliable reading.

Reflecting on the nature of such devilish tradeoffs, Niels Bohr once observed that *contraria sunt complementa* [opposites are complementary]. In particular, he is reported to have said, "Truth and clarity are complementary."[9] That is, truth and clarity exist at the expense of one another. At any given time, we can have some of each — but not ever both completely.

The more we pursue truth — the more closely we scrutinize the intricacies of reality — the less clear things become. Conversely, the more we seek to clarify life's deepest mysteries — the origin of the universe, for instance, or how life came to be — the more we must settle for truths that cannot be proven logically, and the more we must lean on intuition and faith.

In the soft sciences, the loss of certainty has manifested itself in the gradual discovery and documentation of what should have been obvious from the start — that logic isn't very helpful in understanding human behavior. In fact, it can be downright misleading.

I used to author a monthly column for *Psychology Today*, in which I pondered the nuances of human behavior in light of the latest hard scientific research. In one column I featured the outstanding work of Stanford University's mathematical psychologist Amos Tversky. I looked at his lifelong efforts to understand the human decision-making process.

Tversky found that whether it is in selecting a consumer product or voting for a political candidate, we don't make decisions logically. In a paper titled "Rational Choice and the Framing of Decisions," he and coauthor Daniel Kahneman described our decision-making behavior in techno-speak: "The logic of choice does not provide an adequate foundation for a descriptive theory of decision making." What's more, they stated, "The deviations of actual behavior from the normative model are too widespread to be ignored, too systematic to be dismissed as random error, and too fundamental to be accommodated by relaxing the normative system."[10] In plain language: Logic is useless in describing how you and I make many of our decisions. In certain ways, we are fundamentally, arrantly, irrefutably *illogical* creatures.

Many people jump to the conclusion that this fact signifies that something is wrong with us. But the more astute conclusion, as we have seen, is that there is something wrong with logic. It can't even explain arithmetic, so it certainly isn't powerful enough, nor deep enough, to fully explain human behavior.

By the time we are in kindergarten, we discover that logic lacks the ability to explain even the simplest realities of life. Remember? Girls can punch boys and possibly get away with it, but never the other way around. Our older siblings are allowed to say things we aren't. We explain to the school principal that we had nothing to do with the fracas — and we didn't, really — but she doesn't believe us.

As we get older, the unfairness of life — its defiance of logic — doesn't go away. If anything, it only gets worse. If we work hard, we don't necessarily get a promotion. If we treat others well, they don't necessarily treat us well in return. If we eat well and exercise, we don't necessarily stay healthy and prosper.

It is no wonder we are so apt to utter that universal human complaint: "It's not fair!"

Life's frequent and flagrant violation of logic is especially obvious when it comes to love. Consider this shining exemplar of deductive reasoning:

> A equals B.
> B equals C.
> Therefore, A equals C.

Now behold how the *real* world operates:

> Dad loves (daughter) Maria.
> Maria loves Bubba.
> Therefore, Dad loves Bubba — *not!*

The sciences — all three classes of them — have come a very long way since the days of Aristotle, their nominal founder. Back then it really did appear that we could lay our hands not just on truth, but on proof as well. We now know better.

From the queen on down, the sciences have discovered that proof isn't always achievable. They have discovered that we can't ever prove certain truths. Especially — and this is the hardest pill for us to swallow — objective truths about things that really matter to us.

The Bible

During the Middle Ages, it was theology — the study of God and his relationship to us and the universe — that enjoyed the royal title "queen of the sciences." And for good reason. Like mathematics, theology deals with abstractions, with things that aren't necessarily of this world but are enormously relevant to us.

Like God. For most people, God ranks very high on the significance list. According to yearly Gallup polls, for instance, more than 90 percent of Americans believe in God. The vast majority also say their religious beliefs are "very" or "fairly" important to them.[11]

According to a 2011 Reuters/Ipsos poll of people living in twenty-three countries, a majority said "they were convinced there is an afterlife and a divine entity." Only 18 percent said they don't believe in a god, and 17 percent weren't sure.[12] Likewise, according to the Pew Research Center, clear majorities in twenty-two out of thirty-nine nations worldwide say that "it is necessary to believe in God to be moral and have good values."[13]

Theology is like mathematics in another way. In pursuit of certainty, theologians have often resorted to logical proofs.

THE CERTAINTY OF UNCERTAINTY

In *Proslogion*, a lengthy meditation published in 1078, the Benedictine monk Saint Anselm of Canterbury used logic to prove that God exists. His reasoning can be paraphrased this way:

> God is the greatest being imaginable.
> A real being is greater than a hypothetical one.
> Therefore God is real.

Bam!

In Anselm's own words (translated into English): "There is, then, so truly a being than which nothing greater can be conceived to exist, that it cannot even be conceived not to exist; and this being thou art, O Lord, our God."[14]

The problem with any logical proof, however, is that it only works if we buy into the premises on which it is built. In Anselm's proof, the two premises are:

> God is the greatest being imaginable.
> A real being is greater than a hypothetical one.

If we accept the reasonableness of these premises, then Anselm has proven his point. If not, then he hasn't.

Are there viable alternatives to logic? Ways to obtain some degree of certainty that aren't constrained by the delimiting rules of logic? Yes. They are called intuition and faith.

Intuition and faith. What exactly are these two inseparable, companionable abilities?

If we think of an objective truth as a gemstone hidden in a mine, then logic is a well-understood method that walks us to the treasure, one step at a time. By contrast, intuition, less well-understood but extremely powerful, *flies* us there.

Like an airplane, intuition propels the imagination, defies logic and conventional wisdom, and lands us on ideas so novel, so utterly out of this world, that they commonly leave us breathless. Examples of this abound in the wonderful book *They All Laughed* by Ira Flatow.[15] I encourage you to read it.

Above all, read the Bible if you haven't already. Or even if you have already read it. It speaks a great deal about intuition — specifically, about divine intuition, or revelation, which comes directly from God. (For more on this, see chapter 10.)

Where does *faith* fit in?

If intuition or revelation is the airplane that enables us to fly high over logic, then faith is the discipline that makes it possible for us to hang on for the exhilarating ride. Faith is Abraham willing to sacrifice his precious son Isaac even though doing so made no sense. Faith is Moses at the burning bush finally saying yes to God even though it definitely was not in his best interest to do so. Faith is Jesus in the garden of Gethsemane surrendering to God's will even though it meant an agonizing death and made him appear to be an utter failure to his disciples and fellow Jews.

Many people have the mistaken impression that faith is fuzzy-mindedness and, worse, the enemy of reason. But they are wrong.

As the Bible makes clear — and as Gödel's incompleteness theorem and Heisenberg's uncertainty principle affirm — logic isn't powerful enough to make sense of the real world. Intuition and faith pick up where logic and scientific experiments leave off. Intuition and faith are the best-known means we have for obtaining a reliable degree of certainty in matters far too complex for ordinary reasoning to handle.

Truth is bigger than proof.

In the fourth and fifth centuries, Augustine of Hippo elaborated brilliantly on this striking reality. He warned us that logic, notwithstanding its great importance, can reveal only the tip of the iceberg of what there is to be known: "Men go abroad to admire the heights of mountains, the mighty waves of the sea, the broad tides of rivers, the compass of the ocean, and the circuits of the stars, yet pass over the mystery of themselves without a thought."[16]

Using intuition and faith, he explained, we are able to apprehend objective truths that are far beyond the ken of mere logic and sensory perception. "Therefore," he instructed, "seek not to understand so that you may believe, but believe that you may understand; for 'unless you believe, you will not understand.'"[17]

In situations where logic falters, intuition and faith lead the way, like twin guide dogs. To use Anselm's words: In such situations, certainty is best obtained by "*fides quaerens intellectum* — faith in quest of understanding"[18] — not the other way around.

This critical point deserves repeating because it is so foreign to today's secular way of thinking. So here it is again, in a nutshell. If we seek to understand God — hands down, the most complex concept out there — logic alone isn't going to cut it. We need to rely on intuition — above all, divine intuition. And we need to have faith in the fruits of intuition, the way science does in the fruits of logic.

This isn't just a lot of hot air. It is nothing less than the definition of a powerful technique for approaching (if not actually achieving) proof concerning the greatest mysteries of the universe — above all, God. It is a formidable counterpart to the scientific method that I call the "religious method."

Augustine applied this religious method assiduously and ended up acquitting Christianity above all other belief systems, including the Manichaean religion he had practiced as a young man. His dramatic conversion, his arrival at certainty, is memorialized in a massive, thirteen-book tome titled *The Confessions of St. Augustine* — generally considered history's first true autobiography — and his dazzling Christian apologetic *The City of God*. Augustine reportedly rhapsodized, "To fall in love with God is the greatest romance; to seek Him, the greatest adventure; to find Him, the greatest human achievement."[19]

Augustine was not the only religious scholar who, in the pursuit of certainty, successfully married revelation and faith with logic and experience. In the latter centuries of the Middle Ages, he was succeeded by the sages Moses Maimonides and Averroës, each of whom entered into the aforementioned, powerful marriage in order to explicate Judaism and Islam, respectively.

They, in turn, were succeeded by Thomas Aquinas, the incomparable thirteenth-century Dominican intellectual whose premise was that "faith builds on reason, and grace builds on nature." His unprecedented, controversial marriage of Christianity and Aristotelian science and logic — his *Summa Theologica* — ultimately won the blessing of a very skeptical Roman Catholic Church.

The church was initially cynical about Aquinas's undertaking because Aristotle was a pagan, and his philosophy, while not atheistic, was not obviously reconcilable to Christianity's belief in a personal God, among other things. Aquinas not only accomplished the reconciliation masterfully, but along the way also made certain astute corrections to Aristotle's ancient reasoning.

I believe it is safe to say that Aquinas's breathtaking achievement cemented theology's standing as the true queen of the sciences. In grand fashion, he demonstrated the princely roles that revelation and faith play when we take on mysteries far more challenging than just the complex physical processes of the universe. He demonstrated that the religious method — the revolutionary concept of "faith in quest of understanding" — is every bit as potent as the scientific method, which it supplements.

Supplements, not *supplants*.

According to Aquinas, theology is a "sacred science" in which revelation and faith brilliantly complement experimentation and logic: "For the truth about God, such as reason can know it, would only be known by a few, and that after a long time, and with the admixture of many errors; whereas man's whole salvation, which is in God, depends upon the knowledge of this truth.... It was therefore necessary that, besides the philosophical sciences investigated by reason, there should be a sacred science by way of revelation."[20]

Proverbs 3:5 puts it this way: "Trust in the LORD with all your heart and lean not on your own understanding" — wise council seconded by authorities no less formidable than Augustine, Maimonides, Averroës, Aquinas, and so many of our history's great religious sages.

THE CERTAINTY OF UNCERTAINTY

Trust in the Lord.

The scientific method begins with logic and experience, then walks us as far as possible to a place of inevitable uncertainty. By contrast, the religious method starts with divine intuition and faith, then flies us to a place of sublime certainty.

Trust in the Lord.

Our religious search for truth — our sacred quest for God and a personal relationship with him — commences with our acknowledging an inner voice telling us that God exists and continues with our choosing to have confidence in that voice. That is all it takes: a conscious choice to trust in the Lord.

According to the Bible, when we choose to trust in the Lord — when we muster just enough faith to take that first, small step forward — our being breaks free of its worldly moorings and takes flight. From the resulting elevated view, we perceive ourselves and the universe in a radically new way. We begin understanding (albeit imperfectly) things that were incomprehensible with earthbound logic and experience alone.

In the end, if we don't bail out, we are rewarded with a deep-seated sense of certainty, the kind of certainty we all are seeking. The kind that Euclid, Frege, and Russell would envy.

What Does It All Mean to You and Me?

You and I have the ability to prove things we find hard to believe and to believe things we find hard to prove. That is the first and most practical message of this chapter.

I can prove to you, for instance, that a playground can have a limited size without having any boundaries. Sounds impossible, right? It isn't, and I can explain using some elementary topology, the mathematical study of exotic shapes.

Think of a beach ball. Clearly, its surface is limited; it comprises a certain number of square inches. And yet, imagine living on it. You could roam over it freely without ever encountering a single boundary. *Voila!* A finite playground without any borders.

On the flip side of the coin, I can say to you that I believe in the equality of all people — that it is an objective truth. Yet I can't prove it.

In fact, it would be impossible to defend this assertion to someone who insists on being strictly logical or scientific — like James Watson, the American molecular biologist who shared the 1962 Nobel Prize in Physiology or Medicine with Francis Crick for co-discovering the helical structure of DNA. Undeniably a brilliant man.

In 2007 Watson became the target of widespread criticism after he told London's *Sunday Times* that some races, by virtue of certain evolutionary advantages, are genetically superior to others. Regarding Africans, he said, "All our social policies are based on the fact that their intelligence is the same as ours — whereas all the testing says not really."[21]

In 1997 MIT economist Jonathan Gruber used hard-nosed scientific data to defend a similar Darwinian worldview in a paper titled "Abortion Legalization and Child Living Circumstances: Who Is the 'Marginal Child'?" In it he and his coauthors applaud legalized abortion because it helps rid society of riffraff — euphemistically referred to as "marginal" children: "Our estimates imply that the marginal child who was not born due to legalization would have been 70% more likely to live in a single parent family, 40% more likely to live in poverty, 50% more likely to receive welfare, and 35% more likely to die as an infant. These selection effects imply that the legalization of abortion saved the government over $14 billion in welfare expenditures through 1994."[22]

Flying in the face of these brilliant scientists and alleged scientific evidence for the inferiority and disposability of certain persons, even entire races, is an objective truth that lies beyond the comprehension of mere logic and data, a truth that prizes each of us equally, above politics and price tags, beyond individual IQs and socioeconomic status.

> "We hold these truths to be self-evident, that all men are created equal, that they are endowed by their Creator with certain unalienable Rights."

When the Declaration of Independence was drafted and signed, the words "all men" did not include women, children, and people of color. But today the United States of America does operate — or strives to, anyway — on the belief that "all people" — not just white, free-born, property-holding, adult males — are endowed by the Creator with certain unalienable rights. And that this truth is self-evident.

Creator. Self-evident.

In other words, the fundamental equality of all peoples is not the irrefutable conclusion of some Aristotelian syllogism or the outcome of a scientific experiment. It is not a truth I can ever prove logically. It is an intuitive, objective, powerful truth revealed to you and me by our Creator.

Equally, you and I cannot ever prove that the Creator exists — certainly not to everyone's satisfaction. Does that mean the Creator doesn't exist? Of course not. That is the second practical message we should take to heart from this discussion.

Don't let anyone bully you into thinking less of God because you can't "prove" God exists. Likewise, don't be taken in by someone who claims to "prove" that God doesn't exist. As we have seen, science doesn't have the power to prove anything and logic falters when tackling matters more complex than arithmetic.

The third message I hope you will take away from this chapter is this: in today's world, logic and data are vastly overrated, and faith and intuition are vastly underrated.

Some years ago I was at a book-signing event in Los Angeles promoting my book *Can a Smart Person Believe in God?* Among the people who showed up were two individuals I remember quite vividly. The first was a tall man, not well dressed and with a sad demeanor, who explained that he'd been a longtime member of Mensa, the high IQ society. He confessed to me that logic had failed to answer his most important questions about life, and he was hoping that my book might help him.

The second person I remember was a bright high school student. It became clear from our conversation that he had discovered what many older people never get — that when it comes down to it, logic can easily mislead the foolish and unwary. As an example of this, he told me that he and his friends had recently stayed up all night and come up with a logical proof that the universe doesn't exist.

Of course, faith, too, can go awry.

Remember Kurt Gödel, the history-making logician? In his later years, while working at the renowned Institute for Advanced Study in Princeton, New Jersey, he became convinced that someone was out to poison him. He relied entirely on his beloved wife, Adele, to cook his meals and to be his taste tester whenever they were away from home.

In 1977 Adele was hospitalized and could no longer help her eccentric husband. His friends tried everything to get him to eat, but he refused. Eventually the masterful logician succumbed — at the end, weighing just sixty-five pounds. According to the official death certificate, he died of "malnutrition and inanition caused by personality disturbance."[23] In plain language, he starved himself to death.

Faith, like logic, can seriously mislead us. But more than that, Gödel's tragic story illustrates the sheer *power* of faith. Gödel believed so strongly that people were out to poison him that it overwhelmed even his Darwinian will to survive.

My faith in God is so powerful that I gladly put it to the test, and so should you. I listen carefully to other points of view to make certain I'm not ignorant of anything important. When I was a grad student at Cornell — having reached a point in my studies when I was entertaining the possibility of God's existence — I immersed myself in the sacred literature of the world's major religions. I never do anything halfway.

I was fascinated with Buddhism and Hinduism. I was intrigued by Islam. For a long stretch of time, I even went to temple every Friday night with my thesis adviser, who was Jewish. But in the end, like

THE CERTAINTY OF UNCERTAINTY

Augustine, I was gobsmacked by the New Testament — by the God of love, life, and light described therein.

Much of what I read in the New Testament sounded totally illogical. In fact, I often say that if we ever set about creating a religion, we surely wouldn't invent Christianity. Take, for instance, when Jesus says, "Many who are first will be last, and many who are last will be first" (Matt. 19:30). *Huh?* Or when he says, "Blessed are the meek, for they will inherit the earth" (Matt. 5:5). *Huh?* Or when he says, "If someone slaps you on one cheek, turn to them the other also. If someone takes your coat, do not withhold your shirt from them" (Luke 6:29). *Huh? Huh?*

Oddly, it is precisely because the wisdom espoused in the New Testament is not entirely logical that my certainty of its veracity is increased, not decreased.

I feel confident saying so because the Bible and science both agree that the deepest truths — the ones concerning the most complex, most important realities of our existence — lie well beyond the realm of logic. In other words, we expect deep, objective truths not to be entirely logical. We expect them not to make perfect sense. First Corinthians 1:18–19 states it this way: "The message of the cross is foolishness to those who are perishing, but to us who are being saved it is the power of God. For it is written: 'I will destroy the wisdom of the wise; the intelligence of the intelligent I will frustrate.'"

That's the God of the New Testament — foolishness to many who consider themselves wise. That's Jesus. That's Christianity. Truths that are bigger than you and me. Bigger than the universe. Bigger even than proof.

CHAPTER 8

LA VIDA LOCA

CAUSE AND EFFECT CAN BE DISPROPORTIONAL

In a nonlinear system the outcome is often indefinitely, arbitrarily sensitive to tiny changes in the initial condition.

MURRAY GELL-MANN

"Truly I tell you, if you have faith as small as a mustard seed, you can say to this mountain, 'Move from here to there,' and it will move."

MATTHEW 17:20

STRANGE AS IT MIGHT SOUND, like people, universes have personalities. Some are calm and their behavior is fairly predictable. Others are edgy and their behavior is impossible to predict.

Which do you think best describes our world? Calm or edgy?

Three centuries ago science was certain that our universe was cool-headed and completely knowable. The Bible disagreed and painted the picture of a universe that is fundamentally jumpy and unpredictable. Today the two pretty much agree: the Bible came closest to being right.

Years ago, when I was in the Antarctic preparing to make history with my live broadcast to North America for *Nightline* (for more, see chapter 2), I conspired with the head of McMurdo Station, our base of operations, to play a prank on my producer, Rick Wilkinson III. Here is how it went down.

After we spent more than a week filming all over the frozen continent, Rick packed our tapes into a cardboard box and handed it off to a pilot from the New Zealand Air Force, who was supposed to fly it to nearby Christchurch. There he would hand-deliver the important package to our editor, who had just flown in from Washington, DC.

The plan was for the editor to spend several days piecing together portions of the raw video in accordance with my scripts. The results of his labors would constitute the bulk of our history-making show. We were on a very tight schedule. Every second counted. There could be no hitch along the way. There was no Plan B.

Enter my practical joke.

Acting on my instructions, the head of McMurdo Station knocked on the door of the dorm room Rick and I were sharing. The two of us were lounging and chatting — the first breather we'd had since landing on the ice. I answered the door, and the director strode in with a cardboard box that was the spitting image of the one Rick had used for the tapes — down to the shipping labels.

"One of my crew found this lying out on the runway," he said. "It looks like it belongs to you."

I stood there straight-faced, waiting for Rick to leap out of his bunk and begin panicking. The on-time delivery of that package spelled the difference between success and failure. It meant *everything*.

What I hadn't counted on was Rick's super-calm personality. Nothing ruffled him. He was *Mr. Coolio*. Which made him the ideal network television news producer but took the air right out of my practical joke.

"This isn't my package," he said, after inspecting the box. He handed it back to the station director, not so much as raising an eyebrow or his voice. "Good try, guys."

And that was it. He'd turned the joke around on us.

At its core our universe is *not* like Rick. If anything, it has a penchant for *overreacting* when provoked in certain ways — not unlike many other news producers I have worked with, who fly off the handle for the tiniest reason.

The implications of this amazing truth concerning the personality of our universe are huge — in fact, as we are about to see, *doubly* huge. One, because it makes for a world and a life that can drive us crazy. And two, because it makes for a world and a life notable for their endless, surprising possibilities.

Science

Winter 1961.

For Edward Norton Lorenz, professor of meteorology at the Massachusetts Institute of Technology, it was a day that began pretty much like any other. But it wouldn't end that way.

Lorenz's research concerned a belief that philosophers had been kicking around since antiquity, the belief that everything in the universe, including life, was completely fathomable. The belief held that there was a way, not yet discovered, for us to know *everything*, even the future.

For centuries the idea remained nothing more than a pipe dream. But on Christmas Day, 1643, in the remote village of Woolsthorpe, England, Hanna Newton gave birth to a posthumous baby, a male child whose father had died recently. The boy grew up well and for a season was hailed as making the pipe dream finally come true. His name was Isaac.

Isaac Newton's achievement came in the late seventeenth century with twin discoveries that were truly breathtaking: calculus and the laws of physics. At that time, science was still wont to treat the universe like a giant, mechanical clock. Newton's innovations appeared to be precisely what was needed to describe the clock's complex gear works.

The excitement was palpable. At last, with the help of calculus and Newtonian mechanics, science would be able to describe with infinite accuracy the past and present behavior of every object in the universe and even to foretell its future — one hour, one day, or one century beforehand!

It was a heady conceit, a bigheadedness that infected science for the next two centuries.

No one exuded this euphoria with more abandon than Isaac Newton's intellectual successor, the eighteenth-century French physicist and aristocrat Pierre-Simon, Marquis de Laplace. He was convinced that a person equipped with calculus and the laws of physics could theoretically know it *all*:

> An intelligence knowing all the forces acting in nature at a given instant, as well as the momentary positions of all things in the universe,... would be able to comprehend in one single formula the motions of the largest bodies as well as the lightest atoms in the world.... To it nothing would be uncertain, the future as well as the past would be present to its eyes.[1]

Newton's *calculus* implied that the behavior of all entities and processes took place without interruption. That is, the cosmic clockworks never skipped a beat. Newton's *laws of physics* claimed that so long as we knew the initial position and velocity of any object, the object's entire lifetime — every microsecond of it — could be reliably computed. Shoot an arrow into the air at a certain speed and angle and, theoretically, science could predict with irrefutable accuracy its exact trajectory, from start to finish.

Newton's innovations, which promised to work like a crystal ball, encouraged the notion that our universe had a personality like that of my producer, Rick. That it was cool and calm and not given to any surprises. In science-speak, such a universe is called *linear* — which is to say, effects are always in line with causes. Tiny effects invariably result from tiny causes. Large effects invariably result from large causes.

It is like dealing with a level-headed, fair-minded merchant. Want a candy bar? That will be a dollar. Two bars? Two dollars. Three bars? Three dollars. And so on. No ugly bombshells. Such was the idealistic, optimistic state of affairs until the twentieth century. Until Edward Lorenz entered the picture. Until winter 1961.

For as long as he could remember, Lorenz had had a scientific bent. "As a boy I was always interested in doing things with numbers and was also fascinated by changes in the weather."[2]

After serving as a weatherman in the US Army Air Corps during World War II, Lorenz earned advanced degrees in meteorology from MIT. He ended up staying at MIT for the rest of his career, starting as a staff member in 1948 and eventually achieving the rank of full professor. In line with scientists before him, Lorenz labored under the impression that weather — like the universe itself — was complicated but predictable. He hoped that with enough diligence and precision, he could one day do what no one had ever done: predict the weather with unlimited accuracy.

By the time 1961 rolled around, Lorenz felt confident he was well on his way to accomplishing his goal. He was proud to have boiled down his weather simulation program to only a dozen equations. The twelve equations were able to juggle the myriad variables that went into creating weather: air and water temperatures, humidity, air pressure, wind velocity, and so forth. That alone was considered an enviable milestone, given how devilishly complicated weather was believed to be.

But on a fateful winter day in 1961, Lorenz discovered something that would turn out to be exceedingly more consequential than his lifelong goal of accurately predicting the weather. Something that would alter radically and forever the way he and others viewed the personality of our universe.

It began with his deciding to rerun his weather-predicting program to double-check his most recent results. It seemed like a little thing. But as he stood watching the output of his trusty, 740-pound Royal McBee LGP-30 computer, he couldn't believe his eyes. Rather than replicating the previous data set — which is what he expected, given that he hadn't changed any of the meteorological parameters — his program was spitting out results that were outlandishly different.

The inner workings of the bulky LGP-30 relied on 113 vacuum tubes, so Lorenz's first hunch was that the computer had blown a tube. But he checked things out and couldn't find anything wrong. Eventually he came upon a possible culprit. When reentering the data for the second run, he had inadvertently entered numbers that were rounded off. They were extremely close to the first set of numbers but not identical.

If that indeed was the explanation, Lorenz was now more puzzled than ever. Weather was a linear system — or presumed to be — so the tiny discrepancies in the input data should have produced only tiny discrepancies in the output data. But that hadn't happened. Instead, the tiny input changes had produced results vastly different from those of the first run.

In the days, months, and years that followed, Lorenz verified the oddball experience over and over again. Each time he entered numbers only slightly different from the ones before, the resulting weather forecast varied wildly — one might even say randomly — from the one before. Clearly, trivial aberrations in the initial conditions could have unforeseen unbridled consequences.

For Lorenz, the conclusion was unmistakable: at least when it came to weather, our universe behaved nothing at all like my laid-back producer, Rick. Its personality was the antithesis of cool, calm, and predictable. It was positively chaotic, jumpier than a caffeine junkie.

Scientists were slow to react to Lorenz's startling discovery. For the most part, the status quo endured; science continued treating the universe as linear, exactly the way Newton and his successors had done. In 1972, however, at a conference in Washington, DC, of the American Association for the Advancement of Science, Lorenz delivered a talk whose very title made his discovery hard to ignore. It was stated in the form of a provocative question: "Predictability: Does the Flap of a Butterfly's Wings in Brazil Set Off a Tornado in Texas?"

The answer was yes. The slightest disturbance could set the world aflutter. Chaos was real.

From then onward, not only other meteorologists, but scientists of every stripe — biologists, astronomers, physicians, you name it — hopped aboard Lorenz's "chaos" bandwagon. Even those in the humanities joined in, some claiming that Lorenz was not, in fact, the first to document the chaotic predilections of our cosmos.

For centuries, authors worldwide had stated variations of a proverb that illustrates the huge consequences of a seemingly trivial failure. In the June 1758 issue of *Poor Richard's Almanack*, Ben Franklin printed this version of it: "For want of a Nail the Shoe was lost; for want of a Shoe the Horse was lost; and for want of a Horse the Rider was lost; being overtaken and slain by the Enemy, all for want of Care about a Horse-shoe Nail."[3]

During World War II, the following, longer version was reportedly framed and posted on the wall of the Anglo-American Supply Headquarters in London as a warning against slackness:

> For want of a nail the shoe was lost,
> For want of a shoe the horse was lost,
> For want of a horse the rider was lost,
> For want of a rider the battle was lost,
> For want of a battle the kingdom was lost,
> And all for the want of a horseshoe-nail.[4]

In the 1845 short story "The Power of Words," Edgar Allan Poe uses the character Agathos to make this arresting observation about the far-ranging ramifications of small hand gestures:

> You are well aware that, as no thought can perish, so no act is without infinite result. We moved our hands, for example, when we were dwellers on the earth, and, in so doing, gave vibration to the atmosphere which engirdled it. This vibration was indefinitely extended, till it gave impulse to every particle of the earth's air, which thenceforward, *and for ever*, was actuated by the one movement of the hand.[5]

In his 1952 short story "A Sound of Thunder," Ray Bradbury makes a similar and scary point about nature's apparent hypersensitivity to subtle changes. A trio of big-game hunters wishing to bag a T. rex travel more than 60 million years back in time. During their expedition, one of the hunters recklessly steps on a butterfly. Upon returning to the present, they discover that the seemingly minor incident has had huge, catastrophic consequences, including the surprise election of a dictatorial American president.[6]

Truthfully, the rough idea of chaos — of small causes precipitating outsized, seemingly random effects — had been bandied about even in science long before Lorenz brought it to light. As the nineteenth century gave way to the twentieth, the French physicist and philosopher Henri Poincaré, for one, pointed out the potentially erratic behavior of many-body systems — for instance, our own solar system. But his ideas were never enthusiastically pursued.

Illustrious forerunners or no, Edward Lorenz is the person generally and rightfully called the father of chaos theory. Thanks to him, the subject not only has become a robust academic discipline, but has been absorbed into the popular culture — even inspiring the title and story line of the 2004 Ashton Kutcher movie *The Butterfly Effect*.

During his long, fruitful life of ninety years, Lorenz analyzed many different examples of chaotic behavior, not just weather-related ones. In the process, he elucidated three cardinal features of chaos. First, chaos doesn't arise only out of complexity. It arises even out of simplicity. Second, chaos doesn't usually emerge slowly; it leaps out at you like a ghoul in a haunted house on Halloween. Third, chaotic behavior *appears* to be random, but it isn't really.

All three of these characteristics are illustrated by a pot of liquid heating on a stove. First, it is a simple situation that requires far fewer equations to describe than the twelve Lorenz used for the weather. Second, depending on the exact values of certain parameters — the thickness of the liquid, the temperature of the stove, the physical dimensions of the pot — the heated liquid will suddenly go from quiescence to madness. Two adjacent molecules will instantly be swept off in different directions. Third, if we repeat the process, our two adjacent molecules will once again flee from one another. And though the precipitous separation appears random, it isn't. In science-speak, it is random in a deterministic way. Let me explain.

Every time we repeat the heating process, every time the liquid hits boiling temperature, our two adjacent molecules fly off with trajectories that appear random but actually trace out a definite pattern. The pattern is called a *strange attractor*.

Typically, strange attractors look like curvaceous line drawings. They can be quite flamboyant, as you yourself can see by googling "strange attractor." Above all, they are *not* purely random patterns. There is a method, you might say, to the apparent madness of chaos.

As a relatively new field of study, chaos theory is raising more questions than it is answering. Above all, it raises the question of why the universe is like this, why it has this wild kind of personality. Is it an accident or is it purposeful? This is not an altogether scientific question, of course. As we will see in the next section, the Bible has quite a lot to say about it. But it is a question we wouldn't even be asking, never mind taking seriously, if it weren't for Edward Norton Lorenz.

In recognition of his great gift to humanity, Lorenz ended up receiving pretty much every major scientific accolade available to a research meteorologist. In 1975 he was elected to the National Academy of Sciences. In 1983 he shared the Crafoord Prize, conferred annually by the Royal Swedish Academy of Sciences on deserving scientists who don't qualify for Nobel prizes. And in 1991 he received the Kyoto Prize for earth and planetary sciences.

I believe the Kyoto Prize committee captured very well the spirit and magnitude of Lorenz's achievement with its citation, which read:

> He made his boldest scientific achievement in discovering "deterministic chaos," a principle which has profoundly influenced a wide range of basic sciences and brought about one of the most dramatic changes in mankind's view of nature since Sir Isaac Newton.[7]

A dramatic and unexpected change, I would add, entirely befitting an edgy universe given to shocking surprises.

The Bible

Within the opening verses of the Bible, we are given two priceless insights concerning the creation of physical reality. We see: (a) how God operates and (b) what kind of universe he is making. In both cases, *small* reigns supreme.

The Creator at work in Genesis is an imaginative deity who's able to achieve colossal things with small, well-directed actions. Using not many more words than a children's picture book, God speaks into existence the heavens, the earth, and everything we see all around — plus a great deal more that we can't see. (For more on this, see chapter 4.) And in the end, our world reflects that selfsame, divine *hyper efficiency*.

Our world is a place where enormous results can be achieved with small, well-focused efforts. It is able to react quickly, grandly, spectacularly to even the tiniest prompting. Our world is hyperactive, capable of enormous chaos, both good and bad.

In the beginning, the magisterial chaos that God stirs up pleases him. Small gives rise to an entire cosmos that is good — very good. But it doesn't take long — a mere two chapters of the Bible — for the situation to take a catastrophic turn. In the third chapter of Genesis, Adam and Eve do something that on some level is a very small thing. They each taste — merely taste — the fruit of a certain tree.

Granted it is the tree God forbade them to touch, but come on — what child hasn't disobeyed his or her father? If Adam and Eve were my kids, I might sit them down and give them a good scolding, and that's it. But that is not what happens. In the spanking-new paradise that God has created — in a world wired to react spectacularly to the slightest provocation — Adam and Eve's seemingly minor transgression has universal and eternal consequences.

Universal and *eternal*. Life doesn't get any jumpier, any more chaotic than that.

Further into Genesis, we are informed that the chaotic fallout from Adam and Eve's infraction has mushroomed into something truly intolerable. Here is how Genesis 6:5–6 describes it:

> The LORD saw how great the wickedness of the human race had become on the earth, and that every inclination of the thoughts of the human heart was only evil all the time. The LORD regretted that he had made human beings on the earth, and his heart was deeply troubled.

What is God's reaction to this global depravity? Once again, he goes small to achieve a grandiose purpose. He appoints a tiny committee — composed of a nobody named Noah and his ragtag family — to give the entirety of creation a fresh start.

True to the nature of a chaotic universe, the humble obedience of these few people unleash universal, eternal consequences just as Adam and Eve's disobedience did earlier — except, this time, it is for our good. All of creation is eventually renewed. And in Genesis 9:12–15, God promises never again to destroy life with a flood:

> God said, "This is the sign of the covenant I am making between me and you and every living creature with you, a covenant for all generations to come: I have set my rainbow in the clouds.... Whenever I bring clouds over the earth and the rainbow appears in the clouds, I will remember my covenant between me and you and all living creatures of every kind."

In the rest of Genesis, we read one historical report after another where *small* persists in wielding enormous power. The ostensible small difference between the offerings given to God by Cain and Abel explodes into a murderous blow-out between the two brothers. God's simple (albeit astonishing) promise to give Abram and Sarai a child in their old age is really the kick start of a vast family dynasty chosen to represent God on planet Earth. Moses eliciting water from a rock in the wilderness and (according to one popular interpretation) seeming to take credit for the small miracle results in God's denying his entry into the Promised Land. Never mind everything else the long-suffering prophet has done humbly and faithfully on God's behalf. These sorts of things happen only in a universe designed for jaw-dropping surprises. A universe wired for chaos.

Beyond Genesis, throughout the Old Testament, we meet heroes who in one way or another are the least among their contemporaries — Hagar, Joseph, David — yet whose lives end up changing the course of human history. We see the disproportionate power of *small* in the New Testament as well — in the profound value of a widow's mite; in the miracle arising from a boy's small number of loaves and fishes; in the faith of a woman who merely touches the hem of the Messiah's garment.

Above all, this remarkable chaotic trait of God's creation is evident in Mary's pregnancy. In the eyes of the world, Mary is far from being anyone special — just another teenager engaged to a local carpenter. Worse, in the world's estimation, Jesus is but the undistinguished couple's bastard son. The lack of respect he endures from his contemporaries is made clear throughout the Scriptures. It is why Jesus is unable to work miracles in his hometown, whose inhabitants scoff: "Isn't this the carpenter's son? Isn't his mother's name Mary, and aren't his brothers James, Joseph, Simon and Judas?" (Matt. 13:55). It is why Nathanael sneers, "Nazareth! Can anything good come from there?" (John 1:46).

Yet, as memorialized in the early twentieth-century poem "One Solitary Life" by James Allan Francis, we see once again how in this world created by God, small and insignificant can prove to be earth-shattering:

> He was born in an obscure village,
> The child of a peasant woman.
> He grew up in another obscure village
> Where he worked in a carpenter shop....
>
> He never wrote a book.
> He never held an office.
> He never went to college.
> He never visited a big city.
> He never traveled more than two hundred miles
> from the place where he was born.

> He did none of the things
> Usually associated with greatness.
> He had no credentials but himself....
>
> All the armies that have ever marched,
> All the navies that have ever sailed,
> All the parliaments that have ever sat,
> All the kings that ever reigned put together
> Have not affected the life of mankind on earth
> As powerfully as that one solitary life.

God could have created a very different world, one that did not reflect his surprising, infinitely creative nature. A world that was supremely calm, like Rick. One that was always carefully measured in its response to stimuli and stature. One that was entirely equitable, predictable, and logical.

A world like a balloon with unlimited room for growth. Each time we added air to it, it would react the same way: it would merely grow a bit larger. There'd never be any chance — any expectation — of something disproportional or unexpected happening. A boring world.

That is definitely not the world God created. The world he created is primed for amazement. A world known for its chaos, even though the Bible never uses the term. A world whose personality fits to a T the three telltale features of chaos that Lorenz described.

First, chaos can arise in the simplest of settings and from the tiniest change of circumstances. In our world, adding just a little bit of air to an inflated balloon can make it go ... *pop*!

In our world, a mere personal choice can spell the difference between the mundane and the miraculous. No one in the Bible describes this amazing truth more strikingly than Jesus does in Matthew 6:31–33: "So do not worry, saying, 'What shall we eat?' or 'What shall we drink?' or 'What shall we wear?' For the pagans run after all these things, and your heavenly Father knows that you need them. But seek first his kingdom and his righteousness, and all these things will be given to you as well."

In a calm, equitable, logical universe, we'd need to work long and hard to earn good things. But in God's economy, we need only put him first in our lives — that's all — in order to reap everything that is genuinely good in the cosmos. It is akin to the amazing truth revealed to Lorenz on that winter day in 1961 — that a tiny alteration in the weather's initial state can have a titanic impact on its entire future.

Second, chaos can arise in a heartbeat, without warning. Out of the blue, Saul of Tarsus was abruptly converted on his way to Damascus. And one day, when we least expect it, the Son of God will reappear on earth. "Therefore keep watch," Jesus says in Matthew 24:42, "because you do not know on what day your Lord will come."

Third, chaos is not haphazard, even if oftentimes it seems that way. Romans 8:28 puts it this way: "We know that in all things God works for the good of those who love him, who have been called according to his purpose."

In the Bible, this is pointedly illustrated by the long, chaotic, convoluted life of Joseph. In going from being a bullied younger brother to a slave to Pharaoh's right-hand man to the hero of Egypt to the redeemer of his own dysfunctional family, Joseph's experience epitomizes what life can be like in a chaotic universe — seemingly random but actually purposeful.

In short, the world that God created is neither predictable, logical, nor dull. It is a world where zany things happen every day. But according to both science and the Bible, not all the zaniness is random. If we but trust the divinely created process — trust the Lord himself — out of life's seeming madness will materialize the singular masterpiece (think unique strange attractor) that is his purpose for each of us.

What Does It All Mean to You and Me?

Chaos intrudes on our lives even before we are born. That is the first important lesson for you and me.

Within each of our cells is a detailed sketch of who we are genetically — a unique set of forty-six chromosomes. One of them — the tiniest of all, the so-called Y chromosome — has the greatest potential of influencing how we turn out. During conception, if both of our sex chromosomes happen to be Xs, then we are fated genetically to be a female. But if one of them is a Y, then genetically we will become a male. All in all, this is the biological equivalent of the butterfly effect. And that is only the start of life's prenatal chaos.

Notwithstanding our genetically determined gender, our embryonic gonads are sexually *bipotential* — at once female and male. If nothing disrupts the environment of the womb, which under normal circumstances is aflood in female hormones, the embryo will develop into a fetus whose body and brain are feminine.

Starting at around four or five weeks, however, if an embryo is genetically male, the minuscule SRY gene in the tiny Y chromosome abruptly comes to life and asserts itself by expressing a single protein that (among other things) reshapes the embryo's would-be ovaries into testes. The testes, in turn, begin producing a small cocktail of male hormones whose huge, chaotic effect inside the female womb results ultimately in the formation of a male child.

After we are born, it doesn't take long for us to learn firsthand the second lesson of this discussion, namely: the tiniest thing that happens to us can have a gigantic effect on our lives — and not always for the worse.

In *Our Family Tree*, Christopher Radko recalls that after graduating from Columbia University with a degree in English, he didn't know exactly what to do with his life. In 1984, with vague aspirations of being in show business, he took a job in the mailroom of a New York City talent agency. That year he chanced to buy a modern replacement for the old cast-iron stand the family had always used to hold up their beloved Christmas trees. The trees were invariably tall and bedecked with hundreds of vintage, mouth-blown ornaments from Poland.

One evening, shortly after the family had finished hanging the last of the ornaments on a towering twelve-foot-plus tree, one leg of the new stand — just one — faltered and the tree came crashing down, shattering the family's one-of-a-kind keepsakes. Radko recalls that "my grandmother told me I ruined Christmas forever because I caused the tree to fall."[8]

After the holidays, still haunted by the accident, he vacationed in Poland, where a cousin introduced him to a glassblower she'd known since high school. Radko commissioned the artist to create authentic-looking, handmade replicas of his family's shattered heirlooms.

Back home, when his friends and coworkers at the talent agency saw the beautiful facsimiles — which were unlike anything available even in the finest shops of New York City — they started buying them for their own trees. Very quickly, Radko recalls, it was "Good-bye, Hollywood. Hello, Christmas!"[9]

Today Christopher Radko — who loves the celebration of Christmas — runs a thriving, multimillion-dollar business that employs thousands of people. More importantly, it seems clear that he has found his true calling — all, arguably, thanks to one faulty leg support on a tree stand.

In the same year that Radko's tree stand set off a chain of life-altering events, I too had a face-to-face encounter with chaos. It happened at the Smithsonian Institution's National Museum of American History in Washington, DC — at a gala reception following a panel discussion concerning George Orwell's dystopian novel, *1984*.

Noticing that the event's distinguished moderator — Fred Graham, then legal correspondent for CBS News — was standing idly with his female companion, I went over to say hello. When he found out I was a scientist, Graham asked me to settle a dispute he was having with his lady friend. It had to do with the large Foucault pendulum that used to hang in the museum's rotunda.

When I finished giving him my expert opinion on the matter, which evidently he liked very much, he asked if I'd be interested in going into the television news business — reporting on the breakthroughs of science. A few months later, I was christened the science and technology contributor for *CBS Morning News*.

That proved to be the start of a long, successful network television career I never planned on having but that put my scientific training to extraordinarily good use. All because of my response to a seemingly insignificant question during a random encounter with complete strangers!

The third and final lesson for us is this: because we inhabit a world where the least thing can turn our lives upside-down, no one and nothing should be considered insignificant. No one, no matter how humble their origins, should be counted out as being a potential history maker.

Genghis Khan was orphaned as a boy and grew up in the wild, yet he united the Mongols and conquered China. Steve Jobs — conceived out of wedlock by two university students who gave him up for adoption — was a college dropout. Yet he cofounded Apple computers and amassed a personal fortune that at the time of his death in 2011 was worth billions of dollars. J.K. Rowling was a dirt-poor single mom who, because of a single idea she had, became an international bestselling author.

In this unpredictable, illogical, exhilarating world of ours, anyone and everyone matters — exactly what God intended. It is why he made a chaotic universe, I believe — so that none of us would ever feel insignificant. So that none of us would ever feel trapped by our circumstances. So that none of us would ever be without hope for a better future.

Would you have it any other way? Would you give up our world of constant change, innovation, and surprise for absolute peace and predictability? I wouldn't. As put by Henry Adams, the Pulitzer Prize–winning author and scion of the early American Adams family: "Chaos often breeds life, when order breeds habit."[10]

That single observation, for me, is what makes our world so precious. Chaos is many things — scary, thrilling, and usually inexplicable. But above all, it is the crazy stuff that both science and the Bible agree is responsible for life's infinite possibilities.

CHAPTER 9

THE COSMIC GRAPEVINE

INSTANTANEOUS COMMUNICATION IS POSSIBLE

*Anyone who is not shocked by quantum
theory has not understood it.*

NIELS BOHR

*"My sheep listen to my voice; I know
them, and they follow me."*

JOHN 10:27

I EXPECT THAT THE VERY IDEA of instantaneous communication will sound esoteric to most people; it is not part of what usually passes for dinner table conversation. But as we are about to see, for both science and the Bible, it is a topic with a very long history — and it is of enormous consequence to you and me.

Simply put, instantaneous communication means that we are able to hear instantly what someone is saying to us. No matter how widely separated the interlocutors might be — though they be stationed at opposite ends of the universe — there is no time delay between speech and hearing.

When I was at ABC News, I was given the rare privilege of conducting a live conversation with cosmonauts orbiting Earth aboard the Russian space station Mir. Excited as I was about it, I faced two significant problems: the language difference and the expected time lag between every transmission and reception.

The first difficulty was easily resolved by my learning a few words of Russian and the services of an expert simultaneous translator. The second problem was not so readily eliminated. In fact, it was intractable; I simply had to live with it. Please understand something: the delays weren't caused by the distance between the cosmonauts and me — a mere 223 miles — but by all the electronic components and interfaces that composed the complex communications link between us.

The result was that my "conversation" with the Russian space pioneers was awkward at best and stilted at worst. It made for fascinating television, but if such time delays interrupted the flow of our everyday conversations on earth, I'm guessing it would drive us crazy pretty quickly.

This illustration is but a hint of why the subject we are about to explore is so important to both science and the Bible, and ultimately to us. The subject of instantaneous communication is important because it concerns the effectiveness of how we communicate with one another in this world. And it is important because it explains how God himself — who is not of this world — can possibly communicate with us mere mortals.

Science

It is easy for two people close to each other to communicate physically. A touch of the hands. A pat on the back. A hug. Separate them and it is still easy to communicate with sound.

In ancient times, the extended reach of sound — an example of what science now generally calls "action at a distance" — was truly a mystery. Aristotle's best guess was that sound was an air disturbance that traveled like a paper airplane from the transmitter to the receiver. The exact shape of the disturbance determined the actual sound.

Alexander of Aphrodisias, a noted third-century AD philosopher, tweaked Aristotle's idea slightly. As opposed to seeing it as a single, distorted knot of air traveling from A to B, he pictured the disturbance as moving forward by repeatedly imprinting its shape onto the air immediately ahead of it. In other words, sound traveled very much like the leading edge of a wave of toppling dominoes.

Their subtle differences aside, Aristotle's and Alexander's educated guesses implied the same two cardinal properties of sound. First, that it takes sound a measurable amount of time to travel from one place to another. Second, that communicating with sound ultimately relies on direct physical contact of one sort or another — be it a solitary air packet reaching out and tickling our ears or myriad air particles bumping into each other.

Regarding the first property, Isaac Newton was the first to come up with a theoretical estimate for the speed of sound — which we now know was in error by only about 16 percent. Subsequently, a series of clever experiments — the earliest ones clocking the speed of sounds produced by guns and cannons — improved the estimate. Today we know the speed of sound to be 767.7 miles per hour at sea level in dry air at 68°F.

Historically, the second property — of the two, by far the most interesting and consequential — helped raise a generic question about actions at a distance. During the early twentieth-century development of quantum mechanics, the question blew up into a controversy that to this day has science by the throat. Let me explain.

In science-speak, interactions between two objects — for example, a conversation between a husband and wife — divide up into two broad categories: *local* and *nonlocal*. For our purposes, think of the former as involving direct physical contact between the two objects and the latter as not. A handshake is a local interaction — and so is any action at a distance, such as sound, that breaks down into a chain of local interactions.

The late speaker of the US House of Representatives Thomas "Tip" O'Neill is credited with the expression "All politics is local." That is what I'm describing here. A local interaction takes place between two parties that, no matter how great the distance between them — say, a politician in Washington, DC, and a farmer in some tiny town in Central California — an uninterrupted chain of local causes and effects connects the two.

Traditionally, scientists eschewed truly *nonlocal* explanations — ones where no chain of local, physical events between two separated objects could be spelled out. Nonlocal phenomena sounded too outlandish, such as the wave of a fairy godmother's wand supposedly making something happen to someone who is hundreds of miles away.

This prejudice was obvious in the seventeenth century, when scientists were struggling to understand magnetism and gravity. These were seeming actions at a distance that stubbornly defied being explained in terms of conventional, local interactions.

Isaac Newton and his contemporaries were convinced that one day an explanation based on local interactions would be found for gravity. To him, the possibility that gravity — or any action at a distance — could be genuinely nonlocal was "so great an absurdity that I believe no man who has in philosophical matters a competent faculty of thinking can ever fall into it."[1]

For centuries, scientists postulated all sorts of imperceptible, local mechanisms that could possibly explain gravity's cosmic-sized, acquisitive influence. Perhaps gravity was an invisible vortex that sucked everything into itself. Or maybe it comprised invisible waves that tugged at objects, the way breaking waves at the beach pull swimmers toward the sea. Or possibly it consisted of invisible currents of some sort, like rip tides, capable of sweeping objects hither and thither.

Come the early twentieth century, Albert Einstein came up with an idea that topped them all: the general theory of relativity. According to the brainy physicist, the movements of objects we attributed to gravity were actually caused by the invisible *topography of space-time*. Objects rode the hills and valleys of space-time — remaining in direct contact with them — like so many engineless go-carts. (For more on this, see chapter 4.)

Einstein's far-out explanation of gravity has been very successful; it has stood the test of time and countless experiments. Nevertheless, there remain nagging troubles, which scientists seeking a unified field theory — a "theory of everything" — are working hard to resolve.

The biggest problem of all is that general relativity appears to be incompatible with quantum mechanics. It is a situation that reminds me of the story behind Europe's famous Chunnel. Years ago, separate teams of engineers excavated underneath the English Channel — one from the English side, the other from the French. With guidance along the way from GPS sensors, the pair of excavated tunnels met in the middle and were perfectly aligned. Thusly was the Chunnel created.

General relativity and quantum mechanics are like the two tunnels, with a major difference. The former theory attempts an explanation of the universe by describing the behavior of very large-scale objects: planets, stars, galaxies, and the like; the latter, by describing the behavior of very small-scale objects: atoms, electrons, protons, and so forth. But when scientists try grafting the two theories — behold! — they *don't* match up properly.

Science is faced with another great problem. Quantum mechanics appears to have revealed at least one action at a distance that seems to be truly nonlocal. It is a bizarre phenomenon called "quantum entanglement" — which Einstein to his dying day derided as *"spukhafte Fernwirkung."* Roughly translated from German, it means spooky/ghostly/mystical action at a distance.

To grasp the basic idea of quantum entanglement, imagine a radioactive atom that spits out two subatomic particles — say, a pair of electrons. The electrons fly off in opposite directions so that in no time flat they are far apart. Like twins separated at birth.

Further imagine that you are in a position to observe the behavior of Electron A, which scientists commonly call "Alice." In particular, you make a note of the way Alice is spinning. Every known electron spins either clockwise or counterclockwise.

This is where things get spooky/ghostly/mystical.

Without explaining how it is possible, quantum mechanics predicts that at the very instant you observe Alice's spin, Electron B — "Bob" — automatically evidences the opposite spin. There is no time lag; the seeming communication between Alice and Bob is instantaneous.

Now here is the real shocker. This thought experiment actually has been conducted, more than once, in labs all over the world. In each instance, the results confirm what I have just described: quantum particles separated at birth maintain some kind of fundamental connection that gives every appearance of being a spooky/ghostly/mystical instantaneous form of communication.

The first experiment was done in 1972 at the Lawrence Berkeley National Laboratory by American physicist John Clauser and a graduate student named Stuart Freedman. Instead of using electrons and measuring spin, they used photons — particles of light — and measured polarization. But the results were the same: each time they measured Alice's polarization direction, it instantly affected Bob's polarization direction in a predictable way.

Increasing the distance between Alice and Bob appears not to make any difference, as if space disappears or doesn't matter during the interaction. As I write these words, a collaborating group of scientists from the University of Geneva, NASA's Jet Propulsion Laboratory, and the US National Institute of Standards and Technology has set a new distance record for instantaneous quantum entanglement communication: 15.5 miles![2]

In a recent experiment at the University of Science and Technology of China in Shanghai, researchers set out to verify that the communication (or whatever you wish to call it) is indeed instantaneous. According to physicist Juan Yin and colleagues, entangled particles appear to communicate with a speed that is *at least* ten thousand times faster than the speed of light — a result consistent with instantaneity.[3]

This sort of news is sure to make Einstein turn in his grave. A sacrosanct feature of his two theories of relativity is that no object and no information can ever travel faster than the speed of light. Is that sacred principle, in fact, being violated? Or is there some entirely novel mechanism at work here? Stay tuned.

Today quantum entanglement has become something of a household word. Expensive kits are even available for purchase that allow us to witness for ourselves the eerie behavior of entangled photons. One kit costs tens of thousands of dollars.

Recently, *Scientific American* ran a blog titled "How to Build Your Own Quantum Entanglement Experiment." The setup costs only hundreds of dollars.[4]

That's not all. Companies and universities worldwide are in a furious, high-stakes race to use quantum entanglement to create quantum computers and even a quantum internet. The inherent instantaneous machinations of such a computer and internet promise to leave today's supercomputers and information superhighways in the dust.

All this demonstrates that science fact is nearly always stranger than science fiction. "Our imagination is stretched to the utmost," observed the late Nobel Prize–winning physicist Richard Feynman, "not, as in fiction, to imagine things which are not really there, but just to comprehend those things which are there."[5]

Nevertheless, some scientists still side with Einstein — and Newton before him. They scorn the idea of a true, nonlocal phenomenon and continue searching for a localized explanation of quantum entanglement — a chain of conventional, local events that add up to an action at a distance that isn't, after all, spooky, ghostly, or mystical. So far they have failed, and an Irish physicist named John Stewart Bell pretty much proved that they always will fail.

In 1964 Bell published a landmark paper in an obscure scientific journal named *Physics, Physique, Fizika*. In the article, the Irishman derived what is now called Bell's Theorem, which simply put proves that quantum entanglement cannot be explained in terms of local interactions. According to the theorem, quantum entanglement is, indeed, the real McCoy — an action at a distance that is fundamentally, irreducibly nonlocal.[6]

Predictably, die-hard scientists are intent on looking for loopholes in Bell's Theorem. But I see none of them leading to anything other than what the increasing number of experiments appears to show.

Like it or not, science appears to have come upon a phenomenon that defies common sense. An action at a distance that challenges what science once believed was impossible. A spooky/ghostly/mystical mode of communication — a way of connecting widely separated entities — that is consistent with what the Bible has always claimed was quite possible.

The Bible

In the garden of Eden, the Bible reports, Adam and Eve interacted intimately with God. Genesis 3:8 describes one such encounter this way: "Then the man and his wife heard the sound of the LORD God as he was walking in the garden in the cool of the day."

Scholars disagree about whether the original couple walked and talked with a physical or spiritual manifestation of God. But either way, Scripture makes it clear that Adam and Eve spoke with the heavenly Father *directly* — or as science would say, via *local interactions*.

That changed radically after Adam and Eve bit into the forbidden fruit. "Then the LORD said to the woman, 'What is this you have done?'" (Gen. 3:13).

According to Genesis, that was God's last direct, verbal communication with the guilty pair. From then on — because of the estrangement caused by their disobedience — he was forced to communicate with humanity indirectly, using nonlocal means.

I'm reminded of the Cheshire Cat in Lewis Carroll's *Alice's Adventures in Wonderland*. At a certain point in the story, everything about the cat disappears except for his toothy grin. With God after the fall, everything about him disappeared except for his voice.

Even in his so-called face-to-face conversations with Moses, God doesn't reveal his true, unadulterated visage — his "glory," as the Bible calls it. In Exodus 19:19, with God descending on Mount Sinai, we are told that "Moses spoke and the voice of God answered him."

According to Psalm 29:4, "The voice of the LORD is powerful;

the voice of the LORD is majestic." And indeed it is. It is a voice that spoke an entire universe into existence. But let's be honest: hearing a person's voice comes in a clear second place to actually seeing him.

In Exodus 33:18, we can feel Moses' frustration at being so close to God and yet so far from him. "Now show me your glory," he pleads. To which God replies: "I will cause all my goodness to pass in front of you, and I will proclaim my name, the LORD, in your presence. I will have mercy on whom I will have mercy, and I will have compassion on whom I will have compassion. But," he said, "you cannot see my face, for no one may see me and live."

With the birth of Jesus, once again we were able to speak directly with God — although the interaction was still not full blown. Philippians 2:7 explains that even though Jesus was fully God and fully human, in coming to earth, he downsized his visible glory: "he made himself nothing by taking the very nature of a servant, being made in human likeness."

According to the gospel accounts of the transfiguration, Jesus escorted Peter, James, and John to the top of a mountain and allowed them to have a partial glimpse of his glory — what Peter later describes in 2 Peter 1:16 as Jesus' "majesty" (note lowercase "m"). In the very next verse, Peter implies that the majesty he and the two others were allowed to witness was still nothing compared to Jesus' true, unveiled countenance — what the disciple describes as God's "Majestic Glory" (uppercase M and G).

When I was a kid madly in love with science, I learned that looking directly at the sun would blind me. Even looking directly at a solar eclipse was dangerous. A safer way to do it, I found out, was by looking through a piece of smoked glass — like the kind used in a welder's helmet. I imagine that is what happened with Peter, James, and John on that mountaintop. Jesus allowed them to look upon God's Majestic Glory but, for their own safety, only "through a glass darkly."

But even though we must settle for just hearing God's voice, at least the communication is instantaneous. The full significance of that reality is made clear in the Bible's hair-raising account of Abraham's willingness to sacrifice his teenage son Isaac.

Starting in Genesis 22:6, we see the old patriarch obediently doing everything God has asked: building an altar, arranging wood on it for the sacrificial fire, binding his son, and placing him atop the kindling. Then we see Abraham taking his knife in hand and raising it high, ready to strike — when suddenly he hears the voice of God's angel. "Abraham! Abraham!...Do not lay a hand on the boy" (vv. 11–12).

Any time delay whatsoever in Abraham's reception of this urgent (note the exclamation marks), last-minute command would have spelled disaster for the boy and his father. God's message was instantaneous — which enabled him to ascertain just how far Abraham would go to obey him.

Judging from a careful reading of the Bible and my own experience, God's voice can appear to be broadcast from a diversity of sources — even nonhuman objects. One of the most famous examples of this is when God spoke to Moses via a burning bush. In their lengthy back-and-forth, there was no evidence of any time lags between transmission and reception — just as in the cases of Abraham and our quantum acquaintances Alice and Bob.

The seeming oddity of Moses hearkening to a voice emanating from a fiery shrub reminds me of the famous painting *His Master's Voice* by English artist Francis Barraud. It shows a cute, black-eared, white dog named Nipper standing transfixed in front of a wind-up gramophone. When I was growing up, the image was used to advertise RCA products.

The story behind the painting goes like this: Barraud inherited Nipper and the antique phonograph from his deceased brother Mark. He also inherited some recordings of Mark's voice, which Francis played for Nipper. Each time he did, Nipper would cozy up to the gramophone and sit there attentively, just as if he were actually in the presence of his old master.

I imagine it was that way for Moses during that life-changing experience on Mount Horeb in Midian. The voice he heard coming from the flames of the burning bush must have been powerful and majestic — unmistakably God's.

Oftentimes God's instantaneous voice appears to come from within us. That was my experience in September 2006, when my family and I were living on a picturesque, thirty-acre country property in central Massachusetts. One afternoon, while my dog Annie and I were strolling along our property's trails, I began complaining to God about my feelings of emptiness. By the world's standards, I'd had a very successful career; yet I felt that my achievements had for the most part glorified only me, not God.

My prime-time television series for the History Channel had just debuted — on some level, a dream come true — but I was saying to God, "Is this all there's ever going to be in my life? More fame, more fortune, more me?"

That is when I distinctly heard an otherworldly voice broadcasting from within me saying: "Be patient. I'm preparing you for something different." In tandem with that instantaneous aural communication came a text message, if you will. In my mind's eye, I saw a single word in capital letters: CHANGE.

For weeks, whenever I strolled those tranquil pathways with Annie, I'd hear/see those same two messages. And each time, they seemed to be coming from within me, as if my entire body were behaving like a receiver of some sort.

Roughly two months later, in November, I received a phone call out of the blue from Dr. John Templeton, the late president and chairman of the John Templeton Foundation. I don't have the space here to describe everything that transpired, but I will tell you this: that call led to Dr. Templeton's asking me to take charge of a multimillion-dollar, multimedia initiative designed to inspire generosity in people.

The initiative — eventually named Philanthropy Project — gave me the opportunity to do two things I'd never done before: make a movie and create a video-rich website. In the process, it introduced me to a world of truly selfless people — very different from the one to which I'd grown accustomed. All in all, it was a game changer for me.

"Be patient, I'm preparing you for something different."

"CHANGE"

Sometimes God's instantaneous voice appears to come from out of nowhere. It seems that is what Saul of Tarsus experienced while traveling to Damascus to chase down Christians. Here is how he describes it in Acts 26:13–15:

> About noon ... I saw a light from heaven, brighter than the sun, blazing around me and my companions. We all fell to the ground, and I heard a voice saying to me in Aramaic, "Saul, Saul, why do you persecute me? It is hard for you to kick against the goads."
>
> Then I asked, "Who are you, Lord?"
>
> "I am Jesus, whom you are persecuting," the Lord replied.

In instances such as these, God's voice appears to be omnidirectional — to come from everywhere at once. It is as if the very molecules all around us are behaving like loudspeakers, which God is using to communicate with us.

Whichever way God's voice appears to reach us, the mere fact that he and we are able to connect instantaneously is evidence that he and we, like Alice and Bob, are entangled somehow. More precisely, the fact that there is a nonlocal connection between us — a spooky/ghostly/mystical action at a distance — is evidence that God and we were once intimately linked before being separated shortly after the birth of the world.

This remarkable vestigial connection with our Creator suggests to the mind an interesting, modern way of thinking about what it means for him to be omnipresent — a way of seeing omnipresence not as something "outside and everywhere," but "inside and everywhere." A way of seeing God's cosmic-sized grandness in terms of his abiding, quantum-sized connection with us — with every one of our cells.

A hundred years ago, it was likely that most people had only a vague idea of what it meant for God to be omnipresent. Perhaps he could be likened to a winter fog that envelops a beachside community or the fragrant smell that fills a baker's kitchen. But today, in light of what both the Bible and quantum mechanics have to say to us about God and his creation, we have recourse to a more accurate way of picturing God's omnipresence. Today we can say with confidence that God is *everywhere* because he is entangled with us in a most granular way. Because of this entanglement, God has the potential to command every single atom inside and outside of our bodies. And because of God's nonlocal connection with every jot and tittle of the universe, he is able to broadcast his presence to us 24/7, wherever we might be.

The Bible assures us of that in both the Old and New Testaments. Psalm 139:7–10 states it this way:

> Where can I flee from your presence?
> If I go up to the heavens, you are there;
>   if I make my bed in the depths, you are there.
> If I rise on the wings of the dawn,
>   if I settle on the far side of the sea,
> even there your hand will guide me,
>   your right hand will hold me fast.

And Romans 8:38–39 puts it like this:

> For I am convinced that neither death nor life, neither angels nor demons, neither the present nor the future, nor any powers, neither height nor depth, nor anything else in all creation, will be able to separate us from the love of God that is in Christ Jesus our Lord.

Our uncanny connection with God is an amazing truth that logic, faith, and common sense all help us to comprehend. Even though God and we are as separated from one another as Alice and Bob, he has never fully let go of us. He still cares enough to maintain an instantaneous connection with us. He still loves us enough to want us back with him in paradise.

What Does It All Mean to You and Me?

Claiming to hear God's voice doesn't necessarily mean that we are crazy; but if we hear it and ignore it, then we are undeniably, certifiably nuts. That is the primary lesson of what science and the Bible teach us in this discussion.

The lesson, in turn, raises three big questions. First: How can we tell God's voice from others that might be competing with it? Every minute of every day, we are bombarded by a din of voices — from our families, friends, colleagues, or even the many oftentimes conflicting voices inside our own heads — all telling us what we ought to do next.

Maybe we have just been jilted and our best friend is counseling us to get even with the cad. Maybe we have been passed up for a promotion and our gut is telling us to quit and find a different employer. Maybe we feel drawn to a certain profession but our parents are advising against it. Which voice do we heed? Is God using our best friend, our gut, or our parents to communicate instantaneously with us? That is possible, but how can we tell?

In each instance, we must apply at least the following tests: (1) Is the voice being extremely logical? and (2) Is the voice contradicting the Bible?

As to the first test, please note that God's voice is oftentimes remarkably illogical (see chapter 7). Jesus declared that the first among us will be last, that the meek will inherit the earth, and that we should love our enemies. None of this sounds the least bit logical, yet it is God speaking. The *worldly* messages we tend to favor sound more logical, more like: "Hey, there's the best and then there's the rest"; "If someone gets you good, do him one better"; "If it feels great, how bad could it possibly be?"

So when you are being hit all around with conflicting advice, be suspicious of any voice that sounds too logical, too much like what you and I would do in the situation at hand. For as Paul instructs in Romans 12:2, God's will for our lives oftentimes sounds decidedly counterintuitive:

Do not conform to the pattern of this world, but be transformed by the renewing of your mind. Then you will be able to test and approve what God's will is — his good, pleasing and perfect will.

As for the second test, remember that truth never contradicts truth. God will never say anything to us instantaneously that flouts what he has already written to us eternally in the Bible. Therefore we need to know the Bible well. For the better we understand it, the more intimately we will know the heavenly Father and the more proficient we will become at discerning his voice from all others.

A beautiful illustration of this exists in the life of a rookery with tens of thousands of northern fur seals, all of whom are vocalizing at once. The noise is so deafening that by comparison New York City's Grand Central Station at rush hour is as quiet as a church. Yet whenever a mother seal returns from feeding at sea, she can pick out in all that cacophony the tiny voice of her pup — and vice versa, the pup can discern the voice of his or her mom. This extraordinary phenomenon has a single purpose: to reunite parent and child. It is the same with God; he calls out to us — by voice and in the Bible — because he wishes to be reunited with us.

The second big question raised by our main lesson is this: What happens when we don't listen to or obey God's voice?

Being disobedient comes to us naturally. Andy Stanley, the popular founder of North Point Ministries, is fond of explaining how extremely good our inner voices are at misleading us. "We talk ourselves into decisions and then come up with reasons to support these bad decisions."[7]

The Bible gives many examples of such self-deception — Adam and Eve, King Saul, the recalcitrant people of Israel during their forty-year exodus from Egypt. But one of the most concise is the story of Jonah.

At the start of the book named for him, we read about God's voice delivering an unequivocal directive to Jonah: "Go to the great city of Nineveh and preach against it, because its wickedness has come up before me" (1:2). In the next sentence we read about Jonah's response to the instantaneous message: "But Jonah ran away from the LORD and headed for Tarshish [in the opposite direction]" (v. 3).

Why does Jonah cut and run? Because God's order doesn't make any sense to him; it isn't logical. Nineveh is so depraved that Jonah can't wrap his brain around why God can possibly care about it.

Choosing to obey his own, inner voice, Jonah hops aboard a ship to Tarshish, which during its voyage is slammed by a monster storm and nearly capsizes. We see how Jonah's flagrant disobedience places not just his own life in danger, but also the lives of everyone aboard the vessel. It is a dramatic illustration of Proverbs 16:25: "There is a way that appears to be right, but in the end it leads to death."

Have you ever knowingly defied God's voice? Of course you have. Who hasn't? But Jonah's story makes it very clear: we turn a deaf ear to God's often baffling, illogical communications at our own peril. And this brings us, finally, to the third big question raised by this lesson: What happens when we *do* listen and obey God's voice?

Miracles happen.

In Jonah's story, his life is spared — and so is the ship and its crew — whereupon he goes straight to Nineveh. And because of his obedient evangelism there, all 120,000 of Nineveh's lost souls are spared. Although, to the end, Jonah never quite gets why God cares so much for the wayward city.

I wish to share with you one final, particularly dramatic example of what can happen when we listen to and obey God's voice.

In the summer of 1977, I was at Cornell University. South of there, in New York City, a serial killer nicknamed Son of Sam was creating a panic. In August the police apprehended him — but only after he had murdered six and wounded seven innocent young people.

Many years later, David Berkowitz — the killer's real name — explained to the media that when he was in his early twenties, he'd joined a satanic cult and made a pact with the devil. "It was just the stupidest thing I had ever done in my life," he recalled recently. "I just let the devil take control of me."[8]

Around that same time, according to Berkowitz, a demon named Sam began speaking to him. The voice deceived him, he claims, and led him to believe that by shedding innocent blood the devil would set Berkowitz free of the emotional torment and loneliness he had felt nearly all his life.

Berkowitz is now in his early sixties and still serving six consecutive life sentences at Sullivan Correctional Facility in Fallsburg, New York. He has been denied parole more than once and assumes he will never get out.

"Sullivan is a chilling fortress made from Israeli sandstone ... lifeless and indestructible," explains well-known criminologist Scott Bonn, who visited Berkowitz recently. Yet Berkowitz himself, Bonn reports, is a disarmingly changed man, someone who "bounced into the visitation room" and "gave me a firm handshake and then a hug."[9]

As Berkowitz recounts, the seeds of his miraculous transformation began to take root in 1987, when a fellow inmate gave him a Bible and told him that "Jesus Christ loves you, and he's got a plan and a purpose for your life." One evening after that, he explains, while quietly reading the Psalms:

> in the darkness of the cell I got down on my knees, and I just began to cry my heart out to the Lord.... I had so much guilt because I had been such a wicked and vile person. And by my bunk, I just prayed to the Lord and said, Lord, I am so sorry for the life I have lived. Lord, I can't live like this anymore, I am so frustrated I ruined everything. I ruined my life. I ruined other lives. I said God have mercy on my life.... And when I got up off my knees — I don't know how long I was down there, maybe 20 minutes, maybe a half hour — I felt different ... relief, a sense of peace, and even I didn't understand it, but I knew somehow God had heard me, something was going to be different. I went to sleep that night and I slept like a baby.[10]

Today, incredibly, Berkowitz is known as the Son of Hope. He works in the prison chapel as the chaplain's clerk, leads in-prison

Bible study groups, and ministers to hurting people worldwide via correspondence and his own website, ariseandshine.org. He is widely respected by inmates and authorities alike and is credited with saving the lives of individuals who were thinking about committing suicide.

Inevitably, some are doubtful about Berkowitz's apparent miraculous conversion — and he is resigned to that. "I can understand that people in prison, out of prison, can be skeptical. But I have put my faith in Jesus Christ. He has done so much for me. I believe in him, and no matter what people say, I'm going to continue to serve him."[11]

In an uncanny way, Berkowitz's story is very much like the account of Saul of Tarsus — the story of a wanton murderer who finally listened to and obeyed God's voice — who, as a consequence, discovered a whole new purpose in life and was even christened with a new name.

The two stories — and the countless others that happen every day — are strikingly like the phenomenon of resonance. That is where a wineglass is bombarded with sound of just the right frequency; the sound waves transform and ultimately shatter the glass. If you will, the sound finally gets through to the wineglass.

That is what God's voice ultimately did to Berkowitz and Saul of Tarsus. One day God's voice managed to break through their hardened hearts. And the resulting transformation was nothing short of miraculous.

When we disobey God, his voice and ours are dissonant — like a pair of tuning forks vibrating at different frequencies. The dissonance plays out in our lives as stress, fear, and selfishness.

But when we obey God — when his voice gets through to us — it is like what happens when two forks are tuned to the same frequency. Strike one of the forks and the other vibrates *spontaneously* in what science calls "sympathetic vibration."

On a practical level, we vibrate in sympathy with God's voice when we quiet ourselves and listen to and obey God's instantaneous communications. Miracles happen, as we have seen, when we behave in a manner that is consistent with his will for our lives.

Something else happens as well. When our lives truly resonate with God's will, we are no longer merely receivers; we become transmitters as well.

That is what I find most exciting — that when we sojourn through life in harmony with God's will, not only do we do as he wishes, but our voice actually *becomes* God's voice. We speak for him — we proclaim his love for all people, just as Paul did and the Son of Hope is doing — in a dark and noisy world.

CHAPTER 10

BEYOND FLEAS AND GRAPES

HUMANS ARE UNIQUE

*So God created mankind in his own image,
in the image of God he created them;
male and female he created them.*

GENESIS 1:27

*I was much struck how entirely
vague and arbitrary is the distinction
between species and varieties.*

CHARLES DARWIN

"A ROSE BY ANY OTHER NAME WOULD SMELL AS SWEET."

William Shakespeare was many things — above all, a brilliant writer — but clearly he wasn't a scientist. In science, names are all-important. They define strict categories according to which scientists pigeonhole and view everything in creation.

The precise nomenclature scientists use to catalog things is at the very foundation of the scientific method. Yet the classification process is beset with nagging challenges, the biggest one being the arbitrariness of the categories.

When I was a kid, my two sisters and I loved Halloween. After a long night of trick-or-treating, we'd rush home, dump our candy on the living room floor, and sort through the dazzling variety.

Identifying and lumping species — lollipops with lollipops, M&Ms with M&Ms, and so forth — was easy. But when we tried deepening and broadening our taxonomic schemes, it quickly became evident that we faced insurmountable difficulties. My sisters might opt for an organizational structure based on texture, for instance: hard candies in one pile, soft in another. I myself might go with dividing up the candy kingdom according to taste: sweet in one pile, sour in another.

But whichever classification scheme we devised, inevitably some candy would confound us. A Tootsie Pop, for example, with its hard candy shell and soft center: where did it fit into my sisters' scheme? Or Pixy Stix, paper straws filled with sweet and sour candy powder: Where did they fit into my scheme or my sisters', for that matter?

Likewise, despite repeated efforts, scientists since Aristotle have not been able to devise a perfect system for classifying creation. To this day, they don't all agree even on what the largest, most basic categories ought to be.

When I was a schoolboy, my science teacher taught us that the whole of creation could be divvied up into just three distinct kingdoms: plant, animal, and mineral. Nowadays, depending on the country — on the scientific school of thought to which it adheres — kids are taught that there are five or more distinct kingdoms.

Moreover, "kingdoms" have been largely trumped by "domains" as the topmost classification level. The now popular three-domain system (bacteria, archaea, eukaryotes) is based on, among other things, whether the biological cells of a life form have nuclei. Even so, some scientists still favor kingdoms as the topmost levels — or opt for other superlative categories called "dominions" or "empires."

Nowhere are the difficulties of classifying creation more evident than in science's ongoing attempts to figure out how and where humans fit into the great scheme of things.

In the eighteenth century, Carl Linnaeus — the eminent Swedish natural historian credited with founding modern biology's scheme for grouping plants and animals — divided the animal kingdom into six classes. He labeled our species *Homo sapiens*, Latin for "wise man," and lumped us into the class *Mammalia* (Latin for "teat") because newborns feed from their mothers' breasts.

But it soon became obvious that even Linnaeus's classification system — brilliant for its day — was not altogether reliable. For instance, according to his original arrangement, whales were incorrectly lumped in with fishes (they are actually mammals) and rhinoceroses with rodents.

Come the nineteenth century, Charles Darwin — the renowned would-be Anglican priest turned natural historian — amped up the confusion with his theory of natural selection. Suddenly science began seeing the human species as a musical note — a sour note, some claimed — in a hoary tune played by the unpredictable duet of genetic and environmental variations.

Today the vast majority of biologists believe that we are a highly evolved kind of ape. But there are some scientists and many laypersons who do not share that view, who believe that humans do not fit neatly into science's mercurial classification scheme.

Despite the heated disagreement, however, both science and the Bible agree on one inescapable truth: whatever name we go by, however we fit into the scheme of life, whether we evolved or were created fully formed, we *Homo sapiens* are indisputably, magnificently, and vexingly unique.

Science

Picture a university biologist and one of her freshman students coming together for a spirited conversation about what precisely makes humans unique. They settle into chairs in the scientist's office. The wall shelves are crammed with technical books and crowned with varnished, labeled skulls of different animal species.

The pair begins by agreeing that sometimes it is hard to draw a distinction between the behavior of humans and that of other animals. Like all mammals (as Linnaeus first observed), our newborns

BEYOND FLEAS AND GRAPES

are designed to live on mothers' milk. Like all chimpanzees, we make war on one another. Like all bonobos, we grovel and brown-nose. The list of similarities is long and compelling.

"But clearly we're unique because of our genetic superiority," the student says. "After all, we have the biggest, most complex genome in all of creation, right?"

"Actually, we don't," the biologist replies. "We used to believe that human DNA contained a couple million genes. But now that science has the ability to inspect DNA up close, we realize a human has only about 20,000 genes. A slew of other animals and even plants have many more genes than that. The simple grape has 30,000 genes and a water flea has 31,000."

The student frowns. "But there's more to DNA than just genes. It is still true that humans have more DNA than any other living thing on the planet, isn't it?"

The biologist shakes her head. "The actual record holder keeps changing as we sequence the genomes of more and more animals and plants; but right now the largest known DNA belongs to a lowly amoeba named *Chaos*. Its DNA is said to be four hundred times more massive than a human's.

"Some scientists dispute that exact figure, but it is still true that an amoeba holds the record for biggest DNA on the planet. The homely marbled lungfish (*Protopterus aethiopicus*) is the record holder among animals with backbones. And the exquisite Japanese canopy flower (*Paris japonica*) tops the list among plants — with fifty times more DNA than a human.

"Ilia Leitch, a botanist at Jodrell Laboratory in England, said this about the record-setting flower: 'We were astounded when we discovered that this small, stunning plant had such a large genome — it is so large that when stretched out it would be taller than Big Ben [which is 316 feet high].'[1] By comparison, a strand of human DNA is about six feet long."

The student takes a moment to digest this surprising information. "Okay, but genes or no, we're still at the top of the food chain. That makes us the most important, most sovereign creature on earth."

The biologist doesn't want to be disagreeable, so she answers gently, "Well, it's true that humans are an apex predator. No other animal routinely feeds on us, and we alone (with help from ingenious weaponry) are able to kill any other animal on the planet — on land, sea, or air — including ones bigger than us.

"But it's probably more sensible to speak of us as occupying some intermediate position on the food chain because, unarmed, we are easy prey to all sorts of other apex predators, such as lions, tigers, and bears. And also to all manner of microscopic bugs, such as the H1N1 influenza virus that in 1918 wiped out upward of 5 percent of the world's human population.

"In 2013, after evaluating typical human diets, Sylvain Bonhommeau of the French Research Institute for Exploitation of the Sea, along with a number of colleagues, calculated that on a scale of 1 to 5 — with 5 being the very top of the food chain — humans rate an average of 2.21. The people of Burundi scored lowest (2.04) because their diet is mostly plant-based. Icelanders scored highest (2.54) because they typically eat more meat than plants."[2]

The student assumes a doubtful expression. "I heard about that study and think their rating system's a bit artificial. I still buy the argument that, all things considered, we humans are the top dogs in all of creation."

"Suppose I agree," the biologist says, grinning. "It still doesn't make us all-important. Actually, just the opposite." She turns in her chair and points to a framed drawing on the wall of a giant ant standing alongside a relatively small leopard. "My friend E.O. Wilson, the famous Harvard sociobiologist, sketched that. He gave me a copy so I'd never forget a truism he loves sharing, namely, 'It's the little things that rule the world.'"

The student nods unenthusiastically. "Yeah, I've heard him say that."

The biologist raises her right index finger for emphasis. "Ants, for instance. We complain about how they spoil our picnics. But if it weren't for them, we'd be drowning in solid waste. They're the little janitors of creation, always cleaning up after us."

The student chuckles, despite herself. "Yeah, I'll remember that the next time I'm sitting on the grass out on the quad eating my lunch and they suddenly show up."

The biologist continues. "We'd miss ants if they suddenly disappeared. And yet they're at the bottom of the food chain. Who'd miss us humans if *we* suddenly disappeared?"

The student flashes a crooked smile. "I see where you're going with this. The bottom of the food chain is its foundation. So the closer something is to the bottom — like ants — the more important they are to the foundation, the piers of creation."

"Exactly. It's like the game Jenga. You know, where you have a tall stack of interlaid wooden blocks and you have to remove them one at a time without the whole thing collapsing?"

The student nods.

"Nothing happens if you remove the top block — it has zero consequence to the integrity of the stack. But dare to remove a block from way down at the bottom and — *bam!* — the stack won't be able to stand — at least, not for long."

"Sure, I get it," the student says.

"If we humans were suddenly to disappear, everything else in creation would probably cheer and keep on going without so much as a hiccup. Our absence would have zero consequence in the long run — or, at most, a positive one."

The student opens her mouth to take issue with the last remark but changes her mind.

The two spend the next several minutes proposing and eliminating other singular traits that might describe humanity's uniqueness.

Sophisticated societies?

"Bees, ants, and termites also have them," the biologist explains. "Leaf-cutter ants live in incredibly well-run, monarchical, rural communities — arguably better organized and run than many human cities and towns.

"Large soldiers with deadly mandibles defend the colony. Medium-sized forager-excavators dig new tunnels and subterranean rooms. They also carry pieces of leaves back to the underground city — a task comparable to a human hauling 660 pounds while sprinting across hill and dale for many an uninterrupted mile.

"Tiny gardener-nurses chew up the leafy fragments, store away the resulting pulp, and farm it for fungus — the sole food source for the entire community. The gardener-nurses also care for the colony's young, a task shared by small specialists who keep the nests clean at all times."

Clever tools?

"Otters and chimpanzees also make them," the biologist points out. "Years ago Jane Goodall discovered that chimps use long, skinny twigs to fish for termites hidden deep inside tall, inaccessible mounds. Sea otters routinely use rocks, driftwood, and empty bottles to crack open the shells of clams and abalone."

Engineering know-how?

"Beavers also have it," says the biologist, sensitive to the student's growing frustration. "Have you ever studied the intricacies of a beaver dam?"

"Not really," the student murmurs.

"I have, and it's a marvel of civil engineering. Beavers begin by strategically placing large-diameter branches parallel to and against the flow of water. They then interlace the heavy poles with smaller, flexible twigs, which creates an infrastructure that is hard to pull apart. The beavers then pack the empty spaces with stones, mud, and grass. The completed dam is virtually impenetrable. Even a heavy bulldozer has a difficult time tearing one down."

Architectural genius?

"Termites also have it," the biologist says quietly. "Soaring African termite mounds — some as tall as a three-story building — can look quite spectacular on the outside. But their interiors are where the African termites' architectural genius is especially evident. The arrangement of living quarters, nursery chambers, and ventilation shafts is so forward-thinking that African architect Mick Pearce was inspired by it to design Zimbabwe's Eastgate Centre, the nation's largest office and shopping complex, as well as the Council House 2 Building in Melbourne, Australia."[3]

The student speaks with exasperation. "But we humans reason! We alone have intelligence. That is why Linnaeus named our species *Homo sapiens*."

The biologist replies that for a very long time, right up to when she was in high school, that is exactly what science believed. But not anymore.

"There's not universal agreement about what intelligence is, exactly, or how best to measure it; but science now believes that lots of animals have some sort of IQ.

"One proposed indicator of intelligence is the *encephalization quotient*, or EQ, a mathematical comparison of brain weight to body weight. Humans have an EQ of roughly 7.5, placing us at the top of the list. But bottlenose dolphins, orcas, chimps, elephants, dogs, and squirrels are just some of the animals that have EQs greater than 1. In other words, we are exceptional when it comes to intelligence, but not unique."

"Then I give up," says the student, jumping to her feet. "I guess humans aren't so different after all. Certainly, we're not unique."

The biologist, unsmiling, motions her young charge to sit back down. "Sorry to say, you're wrong again."

The student, clearly confused now, hesitates, studies her professor's face for signs of irony. Then she resumes her seat. "Go ahead."

"There *is* something that makes humans unique — that makes us different than any other creature, any other ape, any other humanlike animal alive today, or who's ever lived."

The biologist rises to her feet, walks to one of the shelves, and takes down a skull. It looks human.

"Life on earth has been around for more than three billion years — that's what biologists believe from studying the fossil record.

But never mind that. What's important is that something truly remarkable happened about 50,000 years ago — give or take. Like yesterday in the great scheme of things."

The biologist hesitates.

"What?" the student says, leaning forward. "What happened?"

"Jared Diamond, the celebrated evolutionary biologist, calls it the Great Leap Forward.[4] It's the moment when human beings exploded onto the scene."

The student wrinkles her nose. "You mean *evolved* on the scene."

"That's the party line. But it happened so abruptly, science is in the dark about what actually occurred, and how. Diamond puts it this way: 'Insofar as there was any single moment when we could be said to have become human, it was at the time of this Great Leap Forward 50,000 years ago.... What happened at that magic moment in evolution? What made it possible, and why was it so sudden?... This is a puzzle whose solution is still unknown.'"[5]

The student shifts in her chair. "Why are you telling me this? So what if we showed up suddenly and magically? What does this Great Leap Forward have to do with humans being unique?"

"Everything!" The biologist sets down the skull and looks directly at the student. "Before we — *poof!* — appeared on the planet, no creature had behaved exactly the way we do — created music and art, spoke intelligently, buried their dead with great ceremony, believed in an afterlife.... Hold on, let me show you."

The biologist leaves the room and returns a moment later. "Here, check these out." She gingerly places some delicate bones into the student's cupped hands.

The student holds them up, inspecting them carefully. "What are they?"

"Flutes made from tiny animal bones. Before humans, nothing like them had ever been invented. Same with the cave paintings in France — Lascaux. I've seen them; they're so exquisite I got the chills. Same with the bas-reliefs, necklaces, pendants, sculptures — all made by human hands, all unique to our kind. None of it has any obvious evolutionary advantage, yet all of it is fundamental to being human."

The student hands back the finely carved instruments. "Wow."

"To our knowledge, no other creature on the planet senses the existence of an invisible realm — daily takes into account not just the here-and-now but the hereafter. No other creature believes in a Creator and worships him — builds magnificent cathedrals, composes brilliant symphonies, or founds universities and hospitals in his name.

"In short, we humans are no mere *Homo sapiens*. We're a very special subspecies of anatomically and behaviorally modern humans called *Homo sapiens sapiens*. We're not only wise; we're spiritual. And *that* is what makes us absolutely unique."

The Bible

The Bible divides up creation differently from the scientific explanation, as we will see. But, like science, it sees humans as meriting a special classification — describes us as spiritual creatures who are decidedly unique in the whole of God's ecosystem.

Years ago, I decided to read the Bible cover to cover, no longer willing to settle for being spoon-fed random verses by pastors delivering twenty-minute sermons. This coincided with my marrying Laurel, whom I'd met during my graduate studies at Cornell. Sitting down together, she and I began with Genesis 1:1 and didn't stop until a few years later with the last word of Revelation 22:21.

We didn't speed read our way through the Bible. We were very studious about it, taking on no more than a chapter at a time, filling notebooks with notes and asking a zillion questions along the way.

I remember being struck by the darkness of the Old Testament, where God is nearly always angry, and the relative lightness of the New Testament, where God finally makes peace with us. I also was taken by the Bible's classification system of the earth's plants and animals — a taxonomy that notwithstanding certain obvious differences is comparable to that of science.

In science's cataloging system, plants are said to be *producers* and occupy the lower levels of the food chain. In being able to produce their own foodstuff naturally, via photosynthesis, plants feed not only themselves but also many animals, which are classified as *consumers*.

The Bible sees plants and animals in very much the same light, as producers and consumers. In Genesis 1:29, on the sixth day of creation, God explains this fundamental aspect of the food chain to Adam and Eve. "Then God said, 'I give you every seed-bearing plant on the face of the whole earth and every tree that has fruit with seed in it. They will be yours for food.'"

An argument can also be made that, like modern biologists, the Bible breaks down all living things into three *domains*. The biblical domains are (1) living things without souls, (2) living things with souls, and (3) a single living creature that has both a soul and a spirit.

Plants, I believe, fall naturally into the first domain — soulless entities that have only a physical existence. Please note there is no explicit reference in the Bible to defend this claim, but I argue it is a reasonable conjecture given, as we are about to discuss, the biblical concepts of soul and spirit.

Many people conflate soulfulness and spirituality, but Scripture makes a distinction between the two. This distinction helps us fathom what makes humans truly unique.

My own education concerning the difference between soulfulness and spirituality began shortly after my wedding day, when Laurel and I received a Hebrew-English Torah from our friends Rabbi Leonard Troupp and his wife, Michelle Levine. Even now, twenty-three years later, the large volume occupies a special place on my top bookshelf — and in my memories as well, because it is what got Laurel and me started on our cover-to-cover reading of the Bible.

What I especially recall are the many detailed footnotes and learned commentaries accompanying the Torah's sacred text. In particular, when we read the account of creation, the Torah explained the multiple and subtle meanings of the Hebrew words used to elucidate the difference between soulfulness and spirituality.

One such key word, *nephesh*, can mean any of the following: soul, self, life, creature, person, appetite, mind, living being, desire, emotion, passion. Drill deeper, and it can mean (according to the definitive Brown, Driver, Briggs *Hebrew and English Lexicon*) "that which breathes, the breathing substance or being, soul, the inner being of man."[6]

Another key word is *chayyah*, which means living thing or animal. Throughout the Bible — most notably in its reports about creation and Noah's ark — there appears the pairing *nephesh chayyah*. In most popular versions of the Bible, the English translation of that apposition is "living creature" or "living soul."

In Genesis 1, *nephesh chayyah* is used to describe all the animals living in the water and on the land. In Genesis 2 it is used for Adam. And in Genesis 9 it is used for all the inhabitants of Noah's ark. To wit, Genesis 9:10 says: "and with every living creature (*nephesh chayyah*) that was with you — the birds, the livestock, and all the wild animals, all those that came out of the ark with you — every living creature (*nephesh chayyah*) on earth." So the Bible appears to indicate that not just we humans but all living, breathing animals that possess any sort of mind, personality, and independent will have souls — are "soulish."

Still, the Bible leaves ample room for us to argue about the exact inventory of soulish animals. For instance, should the domain include insects and other invertebrates (spineless animals) — spinal columns being a rough indicator of an advanced nervous system and native intelligence? What about bacteria and other microscopic life — do they have souls too?

The consensus among Judeo-Christian scholars is that the domain of soulish animals comprises vertebrates only. That includes everything from goldfish, lizards, and mice to cats, dogs, and, of course, humans. As for the invertebrates, they most likely fall into the domain of soulless life-forms, along with plants.

What does it mean to have a soul? How significant is it, given that so many creatures have one? Is there only one kind of soul — the same for a dog as for a human? Is the soul immortal, or does it die with the body? The answers to these questions and a plethora of others depend greatly on our interpretation of Scripture.

One thing is certain: the Bible regards humans as more than just soulish, more than just living, breathing creatures with a mind, personality, and free will.

The evidence for this first appears in Genesis 1:26, when God pauses significantly after creating space, time, and the earth and voices a plan for fashioning a creature unlike any that he already has created en masse — unlike all the plants, all the soulless creatures, all the soulish animals.

"Then God said, 'Let us make mankind in our image, in our likeness, so that they may rule over the fish in the sea and the birds in the sky, over the livestock and all the wild animals, and over all the creatures that move along the ground'" (Gen. 1:26). After the magnificent, frenzied spectacle of creation, suddenly we sense a shifting of gears, the coming of a grand finale. God contemplates out loud a creative act that is intimate, personally meaningful, deeply consequential — and which he executes in the very next verse. "So God created mankind in his own image, in the image of God he created them; male and female he created them" (v. 27).

Adam and Eve — in the classification system of science, *Homo sapiens sapiens* — are a breed apart, godlike, holy. We alone, being made in God's image, partake in his essential, spiritual nature. In Psalm 8:3–5 we can feel David's astonishment as he acknowledges humanity's God-given uniqueness:

> When I consider your heavens,
>   the work of your fingers,
> the moon and the stars,
>   which you have set in place,
> what is mankind that you are mindful of them,
>   human beings that you care for them?
> You have made them a little lower than the angels
>   and crowned them with glory and honor.

BEYOND FLEAS AND GRAPES

Elsewhere in the Bible, we are reminded repeatedly of humanity's fundamental differentness. In 1 Corinthians 15:39, for instance, we are informed that we are in a class by ourselves and that invisible though it is to even the world's most powerful microscopes, our special status infuses every one of our biological cells. "Not all flesh is the same: People have one kind of flesh, animals have another, birds another and fish another."

What exactly is our unique status? What does it mean that we are made in the image of God? That we are designed to be a little lower than the angels? That we are crowned with glory and honor? That we have been given the awesome responsibility of being wise stewards of God's universe?

As it happens, the Bible describes our uniqueness in ways that are in line with everything we have discovered scientifically, and then some. First, we are unique in our creativity. Whenever we compose a symphony, paint a masterpiece, invent a better mousetrap, or found a small business, university, or hospital — whenever we innovate, we mirror the stunning imagination of our Creator.

Second, we are unique in our capacity for selflessness. Every time we behave heroically, altruistically, self-sacrificially — in ways that confound Darwin's law of the jungle; whenever we place someone else's interests, even a stranger's, ahead of our own, we mirror the amazing love and mercy of our Creator.

Third, we are unique in being curious about the far-flung universe and our uncanny ability to understand it — even to the point of sensing the existence of something or someone behind it all. It is a far-reaching awareness that I call SQ, or spiritual quotient — a revelatory intelligence far more discerning than logic (see chapter 7).[7]

It is difficult to explain such revelatory intelligence as having arisen accidentally from the provincial, earthbound machinations of natural selection. Rather, it reminds me of what Henry David Thoreau said: "Knowledge does not come to us by details, but in flashes of light from heaven."[8]

How could the mind and senses of a creature allegedly shaped by an earthly environment possibly be expected to comprehend truths about phenomena entirely foreign to its terrestrial experience? How could humans have conceived of black holes, the quantum vacuum, parallel universes? It is because our minds and senses were shaped by the Creator of those phenomena, and as a result we are able to comprehend them. It is because of our SQ, our exclusive, intimate, native connection with God, that we are aware of his presence and are able to have a personal relationship with him.

Without that unique, divine connection to our Creator, without choosing to exercise our God-given SQ, we are nothing but soulish animals. Listen to how this is explained in 1 Corinthians 2:14, one of my favorite verses: "The person without the Spirit does not accept the things that come from the Spirit of God but considers them foolishness, and cannot understand them because they are discerned only through the Spirit."

We *Homo sapiens sapiens* — even atheists, notwithstanding their professed beliefs — are created to be spiritual beings; and by any other name, we'd still be spiritual creatures. Every time we thank God or shake our fists at him or stubbornly deny his existence; every time we marvel at his mercy or wonder why he doesn't spare good people from bad circumstances; every time we offer God's blessings to someone or bury a loved one and wonder if there is, indeed, a life after this one; every time we decry injustice or assert the dignity of humanity and protect the rights of the powerless, we evidence our SQ, our intimate connection with God, and mirror his goodness, aliveness, and unerring sense of perfect justice.

Of course we don't reflect our Creator exactly, any more than children reflect their parents exactly. We aren't God any more than we are our parents. But God shadows everything we think, say, and do — even, or especially, when we are at each other's throats over whether God truly exists or cares.

We are the only part of creation that God declared to be not just "good" but "very good." We are the beloved sons and daughters of the

Creator, fallen but redeemable. We are the reason God became flesh for a time and why he will return one day.

According to the Bible, we are that special to him.

According to the Creator himself, we are that unique.

What Does It All Mean to You and Me?

The search for truth, and the love of it, begins at home — with each of us gazing into a mirror and asking, "Who or what am I, really?" With each of us gazing into our neighbor's eyes and asking, "Who or what are you, really?" With each of us gazing heavenward and asking, "Does the God of the Bible really exist?"

How we answer the questions — how we truly see ourselves, our neighbors, and the God of the Bible — determines how we treat ourselves and others. It determines the way we come down on important, sensitive issues such as euthanasia, pornography, and abortion. Above all, it influences decisively the kind of society we create and defend.

How do *you* come down on the aforementioned questions? Do you see yourself and others as mere animals just one step removed from an ape? And God as a lowbrow, outdated superstition?

Because of the considerable scientific and biblical evidence, some of which I have attempted to explain to you in this chapter and book, I find it difficult if not intellectually dishonest to dismiss God as a figment of the human imagination and our unique standing among living creatures as inconsequential or mere happenstance. In view of the evidence, I see God as the Author of the universe. I see him as the initiator of the Great Leap Forward, which suddenly introduced a species that is clearly distinct from all other life on earth; a species of spiritual creatures a little lower than the angels and one step removed from God.

We make a big fuss these days — and justifiably — about how various groups of historically slighted people are portrayed in advertisements, the news media, and entertainment. Among them are women, Hispanics, African-Americans, Italians — and now increasingly,

Christians. Every year, at gala functions, awards are handed out to those who foster positive images of these groups.

Yes, images matter.

During World War II, Nazis thought of a million and one ways to degrade the popular perception of Jews and other supposed misfits. Specially designed comic books filled with negative, hateful imagery — of Jews as insects carried on the back of the devil — were required reading in Germany's public schools.

With that in mind, read carefully the words of Philip Zimbardo, Stanford University's renowned social psychologist and author of *The Lucifer Effect: Understanding How Good People Turn Evil*: "At the core of evil is the process of dehumanization by which certain other people or collectives of them, are depicted as less than human."[9]

Less than human.

Professor Zimbardo's use of this phrase implies a belief in the very thing we are discussing, that humans are unique. That we are not mere animals. That to portray us as such — as vermin — is not just bad, it's *evil*.

Why is Zimbardo's use of such an extreme word justified here? Because treating people as "less than human" is not just insulting, not just a violation of their civil rights, not just a heinous thing to do — it is far deeper than that. It blasphemes the very one in whose image we are made. It desecrates our fundamental holiness.

There are some who say that humans do not deserve special treatment, that a puppy's life has the same value as a human baby's. That is a popular worldview among animal rights activists.

But such a position is not in line with the scientific and biblical evidence we have discussed in this chapter. Humans don't deserve unique treatment just because we say they do. Humans deserve unique treatment because all the evidence attests that we *are* unique.

Ultimately, that is the purpose of this entire book. To urge you, when faced with complex issues, to consider thoughtfully both the findings of science and the teachings of Scripture. And to help you comprehend the remarkable agreement that exists between the two.

BEYOND FLEAS AND GRAPES

Many people think that believing in the Bible is stupid or that believing in science is heretical. Many believe it is necessary to choose sides — as though we were the children of a contentious divorce or fans of rival teams. But they are greatly mistaken.

In 1986, when I was teaching at Harvard, pro football hall of famer John Madden faced a predicament: his sons played for rival university football teams. Mike was a wide receiver for Harvard; Joe was an offensive tackle for Brown. On Saturday, November 1, of that year, when the two varsity teams squared off in Harvard Stadium, the media wanted to know which team, which son, John would be rooting for. His answer? *Both.* And true to his word, he reportedly sat on one side of the stadium during the first half and the other side during the second.

In the face of any apparent clash on the playing field of ideas between science and the Bible, I recommend we tear a page from John Madden's playbook. I suggest that we begin by treating both sides with the respect they deserve, that they have earned by virtue of their longevity and respective successes over many centuries. As we have seen in this book, science and Scripture each contributes something invaluable to the strenuous game called life — to the debates we will always have concerning reality's most mystifying aspects.

Far too often I hear people disparaging the Bible because it is old and therefore allegedly outdated. Likewise, I hear people putting down science because its conclusions don't always stand the test of time — it seems to these critics that science changes its mind as often as Imelda Marcos changed shoes.

But genuine truth — biblical and scientific — is not a perishable commodity. It neither spoils nor evaporates with time. It endures. The truths I selected for this book are neither too old nor too new to be believed. They are, I submit to you, ageless, trustworthy insights into how the world works and where we fit in.

In closing, I want to point out another parallel between the primary message of this book and the Madden example. In football, at any given time, one team plays offense; the other, defense. Offensive and defensive players approach the game from opposite sides, using different, time-tested, game-winning strategies and tactics. In some ways, science and the Bible are like that. They seek, encounter, and reveal truths about the universe from very different, seemingly opposing angles, using their own distinct, time-tested methodologies. In this book I call them the scientific and religious methods.

At certain times in our history — during the early Middle Ages, for one — the Bible played offense and science played defense. At others — like today — it is science that is on the offensive and the Bible on the defensive. If you look past the shifting roles, however, at the big picture, I believe you will discover what I did, that science and Scripture have both participated significantly in advancing the ball of human understanding. Each in its own way and time has helped us to score objective truths, chief among them, the ten amazing truths in this book.

That, dear reader, is the final, most important message I leave with you.

NOTES

Chapter 1: Best of Both Worlds

1. Lewis Thomas, *Late Night Thoughts on Listening to Mahler's Ninth Symphony* (Boston: G. K. Hall, 1984), 209.

2. Gerald Weissmann, "Science Fraud: from patchwork mouse to patchwork data," *The Journal of the Federation of American Societies for Experimental Biology* 20 (April 2006): 587-90, http://w.astro.berkeley.edu/~kalas/ethics/documents/painted-mouse.pdf.

3. Marcel De Serres, "On the Physical Facts in the Bible Compared with the Discoveries of the Modern Sciences," in *The Edinburgh New Philosophical Journal* 28 (April 1845): 260.

4. T. D. Jakes, *Instinct: The Power to Unleash Your Inborn Drive* (Nashville: FaithWords, 2014).

5. Pew Research Center, "The Global Religious Landscape" (18 December 2012), http://www.pewforum.org/2012/12/18/global-religious-landscape-exec/.

6. Pew Research Center, "Global Christianity—A Report on the Size and Distribution of the World's Christian Population" (19 December 2011), http://www.pewforum.org/2011/12/19/global-christianity-exec/.

7. Namkhai Norbu, *Dream Yoga and the Practice of Natural Light* (Berkeley, CA: Snow Lion, 2002).

8. David Wharton, "Lance Armstrong trying to get lifetime ban lifted, report says," in *Los Angeles Times* (18 March 2015), http://www.latimes.com/sports/sportsnow/la-sp-sn-lance-armstrong-lifetime-ban-20150318-story.html.

9. Danielle Clode, *Continent of Curiosities: A Journey Through Australian Natural History* (Cambridge, UK: Cambridge University Press, 2006), 153.

10. "Jan Ullrich finally admits to using 'treatment,' says it was to level playing field," in *VeloNews* (22 June 2013), http://velonews.competitor.com/2013/06/news/jan-ullrich-finally-admits-to-using-treatment-says-it-was-to-level-playing-field_291690.

Chapter 2: Beyond Circular Reasoning

1. St. Augustine, *Confessions*, Book XI, Chapter XIV.

2. *Blindsight*, directed by Lucy Walker (2006), http://www.imdb.com/title/tt0841084/. See also Lisa Kennedy, "Wider focus makes 'Blindsight' worth a trek," in *The Denver Post* (14 March 2008), http://www.denverpost.com/movies/ci_8552086?source=infinite.

3. Ibid.

4. G. J. Whitrow, *Time in History: Views of Time from Prehistory to the Present Day* (New York: Oxford University Press, 1988), 65.

5. Ibid.

6. Georg Wilhelm Friedrich Hegel, *Lectures on the Philosophy of History, Volume 1*, trans. E. S. Haldane (Lincoln: University of Nebraska Press, 1995), 125-26.

7. Jalâl ad-Dîn Rûmî, "I Died as a Mineral," in *Classical Persian Literature*, ed. Arthur John Arberry (London: Curzon Press, 1994), 241.

8. G. J. Whitrow, *Time in History: Views of Time from Prehistory to the Present Day* (New York: Oxford University Press, 1988), 57.

9. Edwyn Bevan, translator, *Stoicorum Veterum Fragmenta*, II, frag. 625, in *Later Greek Religion* (Darlington, UK: J. M. Dent & Sons, 1927), 30-31.

10. Aristotle, *Meterologica*, Book I, 339b-27.

11. St. Augustine, *The City of God*, Book XII, Chapter XIII, trans. John Baillie in *The Belief in Progress* (New York: Oxford University Press, 1950), 75.

12. *The Works of Francis Bacon*, collected and edited by James Spedding et al. (London, 1857), 292.

13. Alan Lightman, "14 April 1905," in *Einstein's Dreams* (Vancouver, WA: Vintage, 1993).

14. John Baillie, *The Belief in Progress* (Cambridge, UK: Cambridge University Press, 1951), 76.

15. Jonny Diaz, "More Beautiful You." See, for instance, AZLyrics, http://www.azlyrics.com/lyrics/jonnydiaz/morebeautifulyou.html.

NOTES

Chapter 3: I Am Who I Am

1. National Alliance on Mental Health, "Dissociative Identity Disorder," http://www2.nami.org/Content/NavigationMenu/Inform_Yourself/About_Mental_Illness/By_Illness/Dissociative_Identity_Disorder.htm.

2. See, for instance, Stephen T. Thornton and Andrew Rex, *Modern Physics for Scientists and Engineers*, 4th ed. (Belmont, CA: Brooks/Cole, 2013), 1.

3. Otto R. Frisch, *What Little I Remember* (Cambridge, UK: Cambridge University Press, 1979), 95.

4. Albert Einstein, "Concerning an Heuristic Point of View Toward the Emission and Transformation of Light," in *Annalen der Physik* 17 (March 1905), 132; trans. into English, *American Journal of Physics* 33 (5 May 1965): 2, https://people.isy.liu.se/jalar/kurser/QF/references/Einstein1905b.pdf.

5. See Louis-Victor de Broglie, *Recherches sur la théorie des quanta* (PhD thesis, 1924), reprint (Paris: Masson, 1963), preface.

6. Lee Strobel, *The Case for Christ: A Journalist's Personal Investigation of the Evidence for Jesus* (Grand Rapids: Zondervan, 1998), 256.

7. Ibid., 256-57.

8. Mahatmas Gandhi, "What Jesus Means to Me," in *The Modern Review* (October 1941).

9. Christadelphia World Wide, "Our Faith and Beliefs," http://www.christadelphia.org/belief.htm.

10. Christian History Institute, "Council of Nicea," https://www.christianhistoryinstitute.org/study/module/nicea/.

11. Lee Strobel, *The Case for Christmas: A Journalist Investigates the Identity of the Child in the Manger* (Grand Rapids: Zondervan, 1998).

12. Lewis Carroll, *Through the Looking Glass* (CreateSpace, 15 July 2010), chap. 5: "Wool and Water."

13. Max Delbrück, *Mind from Matter? An Essay on Evolutionary Epistemology* (Oxford, UK: Blackwell, 1986), 167.

14. Robert Louis Stevenson, *The Strange Case of Dr. Jekyll and Mr. Hyde* (Oxford, UK: Dover, 1991), chap. 10, paragraph 1.

15. Lee Strobel, *The Case for Christ: A Journalist's Personal Investigation of the Evidence for Jesus* (Grand Rapids: Zondervan, 1998), 257.

16. Ibid., 260.

17. Ibid., 267.

Chapter 4: Seeing in the Dark

1. "Medical Mysteries" series on *Good Morning America* (28 February–1 March 2000), reported by Dr. Michael Guillen, produced by Melissa Dunst.

2. Ibid.

3. Lisa Winter, "Top 10 Unsolved Mysteries of Science," in *IFLScience!* (25 June 2014), http://www.iflscience.com/physics/top-10-unsolved-mysteries-science.

4. Paul G. Hewitt, *Conceptual Physics*, 4th ed. (Boston: Addison-Wesley, 1981), 51.

5. Isaac Newton, *Principia, Volume II: The System of the World*, trans. Andrew Motte, revised by Florian Cajori (Oakland: University of California Press, 1934), 634.

6. Charles W. Misner, Kip S. Thorne, and John Archibald Wheeler, *Gravitation* (San Francisco: W. H. Freeman, 1973).

7. Mario Livio, *Brilliant Blunders: From Darwin to Einstein–Colossal Mistakes by Great Scientists That Changed Our Understanding of Life and the Universe* (New York: Simon & Schuster, 2013). See also Rebecca J. Rosen, "Einstein Likely Never Said One of His Most Oft-Quoted Phrases," in *The Atlantic* (9 August 2013).

8. See, for instance, NASA, "Dark Energy, Dark Matter," http://science.nasa.gov/astrophysics/focus-areas/what-is-dark-energy/.

9. Bertrand Russell, *Russell: The Basic Writings of Bertrand Russell* (Melbourne, Australia: Allen & Unwin, 1961), 348.

10. "Stephen Hawking: There is no heaven; it's a fairy tale," an interview by Ian Sample in *The Guardian* (15 May 2011), http://www.theguardian.com/science/2011/may/15/stephen-hawking-interview-there-is-no-heaven.

11. "The Order of the Phoenix," in *Nylon Guys* magazine (Winter 2008).

12. Ipsos, "Ipsos Global @dvisory: Supreme Being(s), the Afterlife and Evolution" (25 April 2011), http://www.ipsos-na.com/news-polls/pressrelease.aspx?id=5217.

13. Rabbi Howard Jaffe, "In Judaism what is believed to happen to someone after they die?" at ReformJudaism.org; http://www.reformjudaism.org/judaism-what-believed-happen-someone-after-they-die.

14. Ibid.

15. "The Great Afterlife: a debate between Michael Shermer and Deepak Chopra," in *Skeptic*, http://www.skeptic.com/reading_room/the-great-afterlife-debate/. See also Deepak Chopra, *Life After Death: The Burden of Proof* (Harmony, 2006).

16. Ruud Custers and Henk Aarts, "The Unconscious Will: How the Pursuit of Goals Operates Outside of Conscious Awareness," in *Science* 329, no. 5987 (2 July 2010): 47-50.

17. Berkeley Lab News Center, "Dark Energy Fills the Cosmos" (1 June 1999), http://webcache.googleusercontent.com/search?q=cache:OxlldnY2mQgJ:www.lbl.gov/Science-Articles/Archive/dark-energy.html&hl=en&gl=us&strip=1&vwsrc=0. See also Neta A. Bahcall, Jeremiah P. Ostriker, Saul Perlmutter, and Paul J. Steinhardt, "The Cosmic Triangle: Revealing the State of the Universe," in *Science* 284, no. 5419 (28 May 1999): 1481-88.

18. Fred Hoyle, *The Nature of the Universe* (New York: Harper & Brothers, 1950), 124.

19. Fred Hoyle, *The Intelligent Universe* (Austin, TX: Holt, Rinehart, and Winston, 1984), 19.

20. Royal Institution, London, "Evolution from Space," Omni Lecture (12 January 1982). See also Fred Hoyle, *Evolution from Space: A Theory of Cosmic Creationism* (New York: Simon & Schuster, 1982), 27-28.

Chapter 5: Not of This World

1. Galileo Galilei, *Dialogues Concerning Two New Sciences*, trans. Henry Crew and Alfonso de Salvo (New York: Macmillan, 1914), 42.

2. Ibid., 44.

3. Albert Einstein, "Zur Elektrodynamik bewegter Körper," in *Annalen der Physik* 17, no. 10: 891. See also the English translation by George Barker Jeffrey and Wilfrid Perrett, "On the Electrodynamics of Moving Bodies," in *The Principle of Relativity* (London: Methuen and Co. Ltd., 1923).

4. Roger Bannister, *The Four-Minute Mile* (Guilford, CT: Lyons Press, 1955), 1.

5. Chuck Yeager, "Gen. Chuck Yeager Describes How He Broke the Sound Barrier," in *Popular Mechanics* (November 1987), http://www.popularmechanics.com/flight/a4396/1280546/.

6. From the soundtrack of the film *Atomic Physics* (J. Arthur Rank Organization, Ltd., 1948). See The Center for History of Physics, American Institute of Physics, "Einstein Explains the Equivalence of Energy and Matter," https://www.aip.org/history/exhibits/einstein/voice1.htm.

7. Michael F. Holick, "The Vitamin D Epidemic and its Health Consequences," in *The Journal of Nutrition* 135, no. 11 (November 2005): 2739s-2748s, http://www.ncbi.nlm.nih.gov/pubmed/16251641.

8. Sir William Bragg, *The Universe of Light* (London: G. Bell & Sons, 1933), 1.

9. C. J. Jung, *Memories, Dreams, Reflections*, trans. Richard Winston and Clara Winston (New York: Pantheon, 1963), 326.

Chapter 6: An Egg-straordinary Event

1. "Billy Graham," biography, http://www.biography.com/people/billy-graham-9317669.

2. *"Vos calculs sont corrects, mais votre physique est abominable."* See André Deprit, "Monsignor Georges Lemaître," in *The Big Bang and Georges Lemaître: Proceedings of a Symposium in Honour of G. Lemaître Fifty Years After His Initiation of Big-Bang Cosmology*, ed. André Berger (Reidel, 1984), 370.

3. Georges Lemaître, *L'Hypothèse de L'atome Primitive: Essai de Cosmogonie* (Griffon & Dunod), 30 April 1946. See also Georges Lemaître, *The Primeval Atom: An Essay on Cosmogony*, trans. Betty H. Korff and Serge A. Korff (New York: D. Van Nostrand, 1950).

4. Georges Lemaître, "The Evolution of the Universe: Discussion," in *Nature* 128, no. 3234 (24 October 1931): 705.

5. Lemaître, *L'Hypothèse de L'atome Primitive: Essai de Cosmogonie*. See also Lemaître, *The Primeval Atom: An Essay on Cosmogony*.

6. Arthur S. Eddington, "The End of the World: From the Standpoint of Mathematical Physics," in *Nature* 127 (21 March 1931): 450.

7. BBC radio broadcast transcribed in *The Listener* (April 1949). See also "Fred Hoyle: An Online Exhibition," St. John's College, University of

NOTES

Cambridge, http://www.joh.cam.ac.uk/library/special_collections/hoyle/exhibition/radio/.

8. "Stephen Hawking: Science Makes God Unnecessary," on Nick Watt, *ABC News* (7 September 2010), http://abcnews.go.com/GMA/stephen-hawking-science-makes-god-unnecessary/story?id=11571150.

9. "The Primeval Atom Hypothesis and the Problem of Clusters of Galaxies," in *La Structure et l'Evolution de l'Univers*, ed. R. Stoops (Brussels: Coudenberg, 1958), 1-32. See also *Cosmology and Controversy: The Historical Development of Two Theories of the Universe*, trans. Helge Kragh (Princeton, NJ: Princeton University Press, 1996), 60.

10. "Beyond the Big Bang," *The Universe* series (season 1), History Channel, premiered 4 September 2007.

11. Alexander Heidel, *The Babylonian Genesis: The Story of Creation* (Chicago: University of Chicago Press, 1963), 42-43.

12. Ibid., 47.

13. Steven Weinberg, *The First Three Minutes: A Modern View of the Origin of the Universe* (New York: Basic Books, 1993), 154.

14. Carl Sagan, *Cosmos* (New York: Random House, 1980), 193.

15. Steven Weinberg, "A Designer Universe?" in *New York Review of Books* (October 21, 1999).

16. *Money* magazine (January 1999). See also "Sir John Templeton," http://www.sirjohntempleton.org/biography/.

17. "A Few Million Monkeys Randomly Recreate Every Work of Shakespeare," Jesse Anderson Online, http://www.jesse-anderson.com/2011/10/a-few-million-monkeys-randomly-recreate-every-work-of-shakespeare/.

Chapter 7: The Certainty of Uncertainty

1. "Races disagree on impact of Simpson trial: CNN-Time Magazine Poll," *CNN* (6 October 1995), http://www.cnn.com/US/OJ/daily/9510/10-06/poll_race/oj_poll_txt.html.

2. CC/ORC Poll (released Monday, 9 June 2014, at 10 P.M. ET), http://i2.cdn.turner.com/cnn/2014/images/06/09/cnn.orc.poll.oj.simpson.pdf. See also "Majority of African-Americans now say Simpson was guilty,"

CNN (9 June 2014), http://politicalticker.blogs.cnn.com/2014/06/09/majority-of-african-americans-now-say-simpson-was-guilty/.

3. Roxanne Khamsi, "Mathematical proofs getting harder to verify," in *New Scientist* (19 February 2006), https://www.newscientist.com/article/dn8743-mathematical-proofs-getting-harder-to-verify/.

4. Ibid.

5. Eugene Wigner, "The Unreasonable Effectiveness of Mathematics in the Natural Sciences," in *Communications in Pure and Applied Mathematics* 13, no. 1 (February 1960): 14. See also https://www.dartmouth.edu/~matc/MathDrama/reading/Wigner.html.

6. For a more in-depth explanation, see "A Certain Treasure" in Michael Guillen, *Bridges to Infinity: The Human Side of Mathematics* (Los Angeles: Tarcher, 1985), 11-21.

7. Gottlob Frege, *Grundgesetze der Arithmetik* II (Düsseldorf, Germany: Verlag Hermann Pohle, 1903), appendix. See also Gottlob Frege, "Basic Laws of Arithmetic," trans. Philip A. Ebert and Marcus Rossberg (Oxford, UK: Oxford University Press, 2013).

8. Bertrand Russell, "Reflections on My Eightieth Birthday," in *Portraits from Memory and Other Essays* (New York: Simon & Schuster, 1956), 54-55.

9. Inscription on Niels Bohr's coat of arms, which he designed upon being knighted into the Order of the Elephant (1947) and is displayed in Frederiksborg Castle, Denmark. Also see, for example, Ted Bastin and Clive W. Kilmister, "Complementarity and All That," in *Combinatorial Physics* (Singapore: World Scientific Publishing, 1995), 17. Also "Truth," http://www.calpoly.edu/~rbrown/Truth.html.

10. Amos Tversky and Daniel Kahneman, "Rational Choice and the Framing of Decisions," in *The Journal of Business* 59, no. 4, Part 2: The Behavioral Foundations of Economic Theory (October 1986): S252.

11. "More Than 9 in 10 Americans Continue to Believe in God," in *Gallup* (3 June 2011), http://www.gallup.com/poll/147887/americans-continue-believe-god.aspx. Also, "Religion," in *Gallup*, http://www.gallup.com/poll/1690/religion.aspx.

12. Patricia Reaney, "Belief in a supreme being strong worldwide: Reuters/

NOTES

Ipsos poll," in *Reuters* (25 April 2011), http://www.reuters.com/article/2011/04/25/us-beliefs-poll-idUSTRE73O24K20110425.

13. Pew Research Center, "Worldwide, Many See Belief in God as Essential to Morality" (13 March 2014), http://www.pewglobal.org/2014/03/13/worldwide-many-see-belief-in-god-as-essential-to-morality/.

14. Anselm, *Proslogion*, chap. III. See also translation by Thomas Williams (Cambridge, UK: Hackett, 2001).

15. Ira Flatow, *They All Laughed... From Light Bulbs to Lasers: The Fascinating Stories Behind the Great Inventions That Have Changed Our Lives* (reprint, New York: Harper Perennial, 1993).

16. St. Augustine, *Confessions*, Book X, Chapter XV.

17. *The Fathers of the Church: St. Augustine Tractates on the Gospel of John 28-54*, Tractate 29, trans. John W. Rettig (Washington, DC: The Catholic University of America Press, 1993). Augustine is citing Isaiah 7:9.

18. Anselm, *Proslogion*, preface. See also translation by the St. Anselm Institute, http://www.stanselminstitute.org/files/AnselmProslogion.pdf.

19. There appears to be no primary source for this quote, though it is universally credited to St. Augustine. Some claim its actual author is the late Father Raphael Simon (1909-2006), a monk of the Order of Cistercians of the Strict Observance who, like Augustine, was a convert to the faith. See https://fauxtations.wordpress.com/tag/augustine/.

20. *Summa Theologica* (Part I, Question 1, Article 1) in *Basic Writings of Saint Thomas Aquinas*, Volume 1, ed. Anton C. Pegis (Cambridge, UK: Hackett, 1997), 6.

21. Charlotte Hunt-Grubbe, "The elementary DNA of Dr. Watson," *The Sunday Times* (21 October 2007). See http://vitamin-max.com/word/wp-content/uploads/2007/10/watsoninterview.pdf.

22. Jonathan Gruber, Phillip Levine, and Douglas Staiger, "Abortion Legalization and Child Living Circumstances: Who Is the 'Marginal Child?'" Working paper 6034 (National Bureau of Economic Research, May 1997), abstract, http://www.nber.org/papers/w6034.pdf.

23. Solomon Feferman, "Gödel's life and work," in *Kurt Gödel: Collected Works*, Volume 1: Publications 1929-1936, edited by Solomon Feferman et al. (London: Oxford University Press, 2001), 15.

Chapter 8: *La Vida Loca*

1. The full quote in the original French reads: "*Une intelligence qui pour un instant donné connaîtrait toutes les forces dont la nature est animée, et la situation respective des êtres qui la composent, si d'ailleurs elle était assez vaste pour soumettre ces données à l'analyse, embrasserait dans la même formule les mouvemens des plus grands corps de l'univers et ceux du plus léger atome : rien ne serait incertain pour elle, et l'avenir comme le passé, serait présent à ses yeux.*" M. Le Comte Laplace, *Théorie Analytique Des Probabilités* (Paris: Mme. Ve Courcier, 1814), Introduction, ii.

2. "Edward Lorenz, father of chaos theory and butterfly effect, dies at 90," *MIT News* (16 April 2008), http://news.mit.edu/2008/obit-lorenz-0416.

3. Benjamin Franklin, "Poor Richard's Almanack," June 1758, in *The Complete Poor Richard Almanacks*, Volume 2 (Facsimile edition, 1970), 375, 377. See also *Respectfully Quoted: A Dictionary of Quotations*, compiled by the U.S. Library of Congress and James H. Billington (21 January 1970).

4. See, for instance, "The Almanack," http://www.inheritage.org/almanack/mother-goose-migrates-to-america-american-nursery-rhymes.html.

5. Edgar Allen Poe, "The Power of Words," in *The Works of Edgar Allen Poe*, Volume 8 (New York: Frank F. Lovell, 1902), 131-32.

6. Ray Bradbury, "A Sound of Thunder" in *A Sound of Thunder and Other Stories*, (New York: William Morrow, 2005).

7. "Edward Norton Lorenz," Inamori Foundation, http://www.inamori-f.or.jp/laureates/k07_b_edward/ctn_e.html.

8. Melissa Parker, "Christopher Radko Interview: The Glass Menagerie of Christmas and the Artist Behind the Tree," in *Smashing Interviews Magazine* (20 December 2011), http://webcache.googleusercontent.com/search?q=cache:YzIRuSTYto8J:smashinginterviews.com/interviews/business-people/christopher-radko-interview-the-glass-menagerie-of-christmas-and-the-artist-behind-the-tree+&cd=1&hl=en&ct=clnk&gl=us.

9. Elizabeth Large, "Ornaments of Distinction," in *The Baltimore Sun* (18 December 1994), http://articles.baltimoresun.com/1994-12-18/features/1994352183_1_christopher-radko-radko-ornaments-glass-ornaments.

10. Henry Brooks Adams, *The Education of Henry Adams*, ed. Ernest Samuels (Boston: Houghton Mifflin, 1974), Chapter XVI, 249.

NOTES

Chapter 9: The Cosmic Grapevine

1. Sir Isaac Newton, *Principia*, Volume II: The System of the World, trans. Andrew Motte, revised by Florian Cajori (Berkeley: University of California Press, 1966), 634.

2. Kelly Dickerson, "Quantum Teleportation Goes Farthest Distance Yet," in *Discovery News* (10 December 2014), http://news.discovery.com/tech/gear-and-gadgets/quantum-teleportation-goes-farthest-distance-yet-141210.htm.

3. Brian Dodson, "Quantum 'spooky action at a distance' travels at least 10,000 times faster than light," in *gizmag* (10 March 2013), http://www.gizmag.com/quantum-entanglement-speed-10000-faster-light/26587/.

4. George Musser, "How To Build Your Own Quantum Entanglement Experiment, Part 1 (of 2)," in *Scientific American* (8 February 2013), http://blogs.scientificamerican.com/critical-opalescence/how-to-build-your-own-quantum-entanglement-experiment-part-1-of-2/.

5. Richard Feynman, "Probability and Uncertainty—the Quantum Mechanical View of Nature," in *The Character of Physical Law (Messenger Lectures, 1964)*, chap. 6 (Cambridge, MA: The MIT Press, 2001), 127-28.

6. See, for example, John S. Bell, "On the Einstein Podolsky Rosen Paradox," in *Physics* 1, no. 3 (1964): 195-200. See also http://www.drchinese.com/David/Bell_Compact.pdf.

7. Michelle A. Vu, "Interview: Andy Stanley on the Principle of the Path," in *The Christian Post* (18 April 2009), http://www.christianpost.com/news/interview-andy-stanley-on-the-principle-of-the-path-38158/.

8. Scott Ross, "Interview: Son of Sam Becomes Son of Hope," on *The 700 Club*, http://www.cbn.com/700club/scottross/interviews/SonofSam.aspx.

9. Dr. Scott Bonn, "Serial Killer David Berkowitz, aka Son of Sam, tells professor 'I was once an evil person' in prison conversation," on *CBS News* (18 March 2013), http://www.cbsnews.com/news/serial-killer-david-berkowitz-aka-son-of-sam-tells-professor-i-was-once-an-evil-person-in-prison-conversation/.

10. Ross, "Interview: Son of Sam Becomes Son of Hope."

11. Ibid.

Chapter 10: Beyond Fleas and Grapes

1. "Rare Japanese plant has largest genome known to science," in *Science Daily* (7 October 2010), http://www.sciencedaily.com/releases/2010/10/101007120641.htm.

2. Joseph Stromberg, "Where Do Humans Really Rank on the Food Chain?" at *Smithsonian.com* (2 December 2013), http://www.smithsonianmag.com/science-nature/where-do-humans-really-rank-on-the-food-chain-180948053/?no-ist.

3. Tom McKeag, "How Termites Inspired Mick Pearce's Green Buildings," at *GreenBiz* (2 September 2009), http://www.greenbiz.com/blog/2009/09/02/how-termites-inspired-mick-pearces-green-buildings.

4. Jared Diamond, "The Great Leap Forward: Dawn of the Human Race," in *Discover Magazine* (May 1989). See also http://wps.pearsoncustom.com/wps/media/objects/6904/7070246/SOC250_Ch01.pdf. See also Jared Diamond, *The Third Chimpanzee*, Part 1, Chap. 2 (New York: Harper Perennial, 2006), 32-58.

5. Ibid.

6. Francis Brown, S. R. Driver, and Charles A. Briggs, *The Brown-Driver-Briggs Hebrew and English Lexicon* (New York: Hendrickson, 1994).

7. For more background and information about SQ, see Michael Guillen, *Can a Smart Person Believe in God?* (Nashville: Thomas Nelson, 2006).

8. Henry David Thoreau, *Life Without Principle*.

9. Philip Zimbardo, "Dehumanization," http://www.lucifereffect.com/dehumanization.htm. See also Philip Zimbardo, *The Lucifer Effect: Understanding How Good People Turn Evil* (New York: Random House, 2008).

SCRIPTURE INDEX

Genesis
1	165
1:1	77, 163
1:3	81
1:17, 26	166
1:27	154, 164
1:29	164
2	165
3:8	142
3:13	142
4:14	17
6:5–6	127–28
9	165
9:10	165
9:12–15	128
22:6, 11–12	144

Exodus
3:14	78
19:19	142
33:18	143

Psalms
8:3–5	166
8:5	79
25:2	100
29:4	142
46:10	99
90:2	77
139:7–10	147
139:14	36, 69

Proverbs
3:5	112
16:25	150

Ecclesiastes
3:17	22
7:15	22

Isaiah
45:7	78
55:8	67

Jonah
1:2, 3	149–50

Micah
5:2	29

Matthew
5:5	117
6:31–33	130
13:55	129
17:20	118
19:30	117
24:36	30
24:42	131

Mark
2:5–7	48
8:27	44
14:36	48

Luke
2:10–11	80
3:23	28
6:29	117
24:5–6	61

John
1:3	84
1:46	129

Reference	Page
3:16	18
3:19–21	81
8:12	76
10:27	135
10:30	47
11:35	48
14:1–7	20
14:6	47
16:33	83
20:26, 27	64

Acts
Reference	Page
26:13–15	146

Romans
Reference	Page
1:20	82
8:28	131
8:38–39	147
12:2	148

1 Corinthians
Reference	Page
1:18–19	117
1:20, 27	49
2:14	168
15:39	167
15:42–44	64
15:51–52	64

2 Corinthians
Reference	Page
4:18	52

Ephesians
Reference	Page
2:8–9	37
5:8	83
6:12	65

Philippians
Reference	Page
2:5–7	38
2:7	143

2 Thessalonians
Reference	Page
2:8	79

Hebrews
Reference	Page
13:8	24

1 Peter
Reference	Page
1:24–25	25
3:18	29

2 Peter
Reference	Page
1:16	143

1 John
Reference	Page
1:5	70, 76

Revelation
Reference	Page
1:8	77
11:1–2, 3, 7–9, 11–12	29
21:1–4	30
22:21	163

NAME INDEX

Abel. *See* Cain and Abel
Abraham, 110, 144
Abram, 128
Adam and Eve, 17, 35, 78, 127–28, 142, 149, 164, 165, 166
Adams, Henry, 134
Alexander of Aphrodisias, 136–37
Anderson, Jesse, 98
Anselm of Canterbury, 109, 111
Aquinas, Thomas, 112
Aristotle, 13, 14, 32, 103, 108, 112, 136
Arius, 47
Armstrong, Lance, 21, 22
Augustine of Hippo, 25, 32–33, 110–11, 112, 117
Aurelius Augustinus. *See* Augustine of Hippo
Averroës, 111, 112
Bacon, Francis, 33–34
Baillie, John, 35
Bannister, Roger, 73–74
Barraud, Francis, 144
Bell, John Stewart, 141–42
Berkowitz, David, 150–52
Bohr, Niels, 38, 41, 49, 106, 135, 180–9
Bondi, Hermann, 88
Bonhommeau, Sylvain, 158
Bonn, Scott, 151
Bradbury, Ray, 124–25
Bragg, William, 70, 82
Brown, Nicole, 100
Buddha, Guatama, 26, 44
Cain and Abel, 17, 128
Carroll, Lewis, 49, 142

Carson, Donald, 48
Chopra, Deepak, 63
Clauser, John, 140
Clode, Danielle, 22
Constantine the Great, 46–47
Coulomb, Charles Augustin de, 15
Cox, Brian, 54
Crick, Francis, 114
Custers, Ruud, 65
Darwin, Charles, 16, 154, 156
David, 129, 166
de Broglie, Louis-Victor, 43
Devlin, Keith, 102
Diamond, Jared, 162
Diaz, Jonny, 37
Dirac, Paul, 55
Eddington, Arthur, 57, 88
Einstein, Albert, 11, 14–15, 33, 42–43, 55–56, 57–58, 59, 60, 62, 72–73, 74, 75, 79, 86–87, 105, 138–39, 140, 141
Epicurus, 71
Euclid, 14, 103–4
Eusebius Pamphilus, 47
Eve. *See* Adam and Eve
Fermat, Pierce de, 101
Feynman, Richard, 141
Flatow, Ira, 109
Francis, James Allan, 129–30
Frank, Erich, 27
Franklin, Ben, 124
Freedman, Stuart, 140
Frege, Gottlob, 104
Friedman, Alexander, 86–90
Galilei, Galileo, 71–72
Gamow, George, 89

Gandhi, Mahatma, 46
Gardner, Martin, 103
Gell-Mann, Murray, 118
Genghis Khan, 134
God. *See* Subject index
Gödel, Kurt, 104, 110, 116
Gold, Thomas, 88
Goldman, Ronald, 100
Graham, Billy, 85
Graham, Fred, 133
Greene, Brian, 91
Gruber, Jonathan, 114
Hagar, 129
Hawking, Stephen, 41, 61, 90
Hegel, Georg, 27
Heisenberg, Werner, 105–6, 110
Heraclitus, 31
Herschel, John, 16
Hertz, Heinrich, 41–42
Holick, Michael F., 81
Hoyle, Fred, 68–69, 88, 89
Hubble, Edwin, 24, 52, 57–58, 87, 88
Hugo, Victor, 19
Jaffe, Howard, 62
Jakes, T. D., 18
Jesus. *See* Subject index
Jobs, Steve, 134
John (apostle), 29
John Paul II (Pope), 11
Jonah, 149–50
Joseph, 129, 131
Jung, Carl, 83
Kahneman, Daniel, 107
Kettering, Charles F., 100
Keyes, Daniel, 39
Kipling, Rudyard, 41
Kurt Gödel, 104–5
Lapides, Louis, 45, 50–51

Laplace, Pierre Simon, Marquis de, 121
Lavoisier, Antoine-Laurent, 54
Lazarus, 48
Leitch, Ilia, 157
Lemaître, Georges, 86–90, 96
Lenard, Philipp, 42
Levine, Michelle, 164
Lightman, Alan, 35
Linnaeus, Carl, 155–56, 161
Livio, Mario, 58
Lorenz, Edward Norton, 120, 122–23, 125, 126
Madden, John, 171
Maimonides, Moses, 111, 112
Mani, 78
Mary (mother of Jesus), 18, 28, 129
Michelson, Albert, 54
Morley, Edward, 54
Moses, 44, 77–78, 110, 128, 142–43, 144–45
Muhammad, 44, 46
Murray, Bill, 25
Nathanael, 129
Nemesius of Emesa, 31
Newton, Isaac, 15, 55, 57, 120–21, 137, 138
Noah, 128, 165
Norbu, Namkhai, 20
Oort, Jan, 58
Orwell, George, 133
Pearce, Mick, 161
Perlmutter, Saul, 67–68
Phoenix, Joaquin, 61
Pius XII (Pope), 90, 96
Planck, Max, 15
Plato, 13, 31
Poe, Edgar Allan, 124
Poincaré, Henri, 125

NAME INDEX

Pontius Pilate, 11, 28
Priestley, Joseph, 54
Pythagoras, 14, 40
Radko, Christopher, 132–33
Rømer, Ole, 72
Rowling, J.K., 134
Rûmî, Jalâl ad-Dîn, 28
Russell, Bertrand, 61, 104–5, 113
Sagan, Carl, 95
Sarai, 128
Saul (King), 149
Saul of Tarsus, 131, 146, 152
Schuller, Robert H., 98
Schwarzschild, Karl, 57
Simpson, O.J., 100–101
Socrates, 13
Solomon, 22
Son of Sam, 150–51
Spurgeon, Charles, 19
Stanley, Andy, 149
Stevenson, Robert Louis, 39, 50
Strobel, Lee, 48
Summerlin, William, 13

Templeton, John Marks, 97, 145
Thales of Miletus, 13
Thomas, Lewis, 13
Thomson, William, 41
Thoreau, Henry David, 167
Troupp, Leonard, 164
Tversky, Amos, 106–7
Ullrich, Jan, 22–23
Watson, James, 114
Weber, Dave, 53
Wegener, Alfred, 22
Weinberg, Steven, 84, 94, 96
Wells, H. G., 34–35
Whitrow, G. J., 27, 29
Wigner, Eugene, 102–3
Wiles, Andrew, 101–2
Wilkinson, Rick III, 36, 119
Wilson, E.O., 158
Yeager, Chuck, 74
Yin, Juan, 140
Zarathustra, 78
Zimbardo, Philip, 170
Zwicky, Fritz, 59

SUBJECT INDEX

abiogenesis, 68
abortion, 114
absolute meaning, 12
accident, 68–69
achromatopsia, 53
afterlife, 61–64, 80
agnosticism, 95, 96
air, 54
Alice's Adventures in Wonderland (Carroll), 142
animal kingdom, 155–56, 165
antiparticles, 76
Apostles' Creed, 63
arithmetic, 104
astronomical calendar, 31
astrophysics, 55, 58–60
atheism, 91, 96
atheists, 68, 90, 91, 94, 96, 112, 168
atom bomb, 75
atoms, 41–42

Babylonian gods, 91–92
Ballad of East and West, The (Kipling), 41
behavior, human, 106–7
Belief in Progress, The (Baillie), 35
Bell's Theorem, 141
Bible
    cause and effect, 127–31
    certainty and uncertainty, 108–13
    classification systems, 163
    and contradictions, 44–49
    and hidden reality, 60–64
    and humans' uniqueness, 163–69
    and instant communication, 142–48
    and light, 76–80
    and linear time, 27–30
    and objective truth, 17–21
    and origin of the universe, 91–96
big bang hypothesis, 33, 68, 89, 91
biology, 14
black holes, 58, 59
blindness, 53
Blindsight (documentary), 26
brightness, 43
Brilliant Blunders (Livio), 58
Buddhism, 18, 26, 37, 46, 62
burning bush, 49, 77, 110, 144–45

calculus, 120, 121
Can a Smart Person Believe in God? (Guillen), 115
Case for Christ, The (Strobel), 50
Case for Christmas, The (Strobel), 48
cause and effect, 118–34. *See also* chaos
    and the Bible, 127–31
    and science, 120–26
    what it means in everyday life, 131–34
celestial ghosts, 57
certainty and uncertainty, 100–117. *See also* faith
    and the Bible, 108–13
    and science, 102–8
    what it means in everyday life, 113–17

SUBJECT INDEX

chaos, 124, 125–26, 127–28, 130, 131–32, 134. *See also* cause and effect
chayyah, 165
Christadelphians, 46
Christianity
    and afterlife, 61–62, 62–63
    and linear time, 27, 28–30, 33
    and objective truth, 20
    world's most practiced religion, 19–20
"Christian Miracle," 18, 28
Christmas, 79–80
chromosomes, 132
Circle of Life, 26, 27–28
circular time, 25–26, 27–28, 30, 31–32, 35
City of God (Augustine of Hippo), 33, 111
civic calendar, 31
clarity, 106
classification systems, 154–56, 163, 164
color, 43
communication, instantaneous. *See* instantaneous communication
complexity, 125
Confessions of Saint Augustine, The (Augustine), 111
conscious will, 65
constants, physical, 15
consumers, 163, 164
continental drift, 22
Continent of Curiosities (Clode), 22
contradictory nature, 38–51, 149
    and the Bible, 44–49
    practical application to everyday life, 49–51
    and science, 40–44

cosmic egg theory, 88, 89, 90
cosmos, 14, 18, 34, 73, 82, 85, 89, 91, 93, 95, 124, 127, 131. *See also* universe
Council of Nicaea, 47
creation, 49, 91–96, 163, 166
    classifying, 155
    goodness of, 168–69
    moment of, 88
    mysteries of, 49
creativity, 98, 167
Creator, 37, 115, 127, 147, 163
Crystal Cathedral, 98
curiosity, 167

dark energy, 34, 60, 67
dark matter, 59, 67
darkness, 79, 163. *See also* hidden reality; light
Da Vinci Code, The (movie), 45
death, 60–61, 80. *See also* afterlife
decision-making, 106–7
deductive reasoning, 103, 107
dehumanization, 170
depravity, global, 128
design, intelligent, 69
deterministic chaos, 126
devil. *See* Satan
dharmas, 28
disobedience, 149–50
dissociative identity disorder (DID), 39
dissonance, 152
divine intuition, 111
divine revelation, 110, 112
DNA, 114, 157
domains, 155, 164
Dr. Jekyll and Mr. Hyde (Stevenson), 39, 50

dualism, 63, 78–79

Easter, 47, 63
Eastern religions, 20, 26, 62
Eden, 142
Egypt, ancient, 30–31, 33
Einstein's Dreams (Lightman), 35
electrons, 42, 44, 76, 139
Elegant Universe, The (Greene), 91
encephalization quotient (EQ), 161
energy, 75–76
Enuma Elish, 91–92
equality, 115
ether, 54, 60
evil, 78–79, 81, 82, 170
evolution, 162
exceptionalism, 95, 97
experimental method, 34
extraterrestrial beings, 70–71

faith, 67–68, 102, 105, 106, 109–13, 115, 116, 148, 152. *See also* certainty and uncertainty
Fermat's Last Theorem, 101–2
final judgment, 30
fire, 53–54
flood, 128
food chain, 157–59, 163, 164
forgiveness, 48, 81
free will, 82

galaxies, 56, 58–59, 68, 87, 88, 93, 95, 139
genes, 157
geometry, 14, 104
Gnosticism, 46, 47
God
    and Adam and Eve, 142, 164
    as Author of universe, 169
    belief in, 108–9
    beloved children of, 168–69
    conversation with Moses, 143
    existence of, 16, 96, 109
    and forgiveness, 81
    image of, 37, 166, 167
    instantaneous communication with, 144, 145, 152
    and light, 76–80, 81–82
    personal relationship with, 168
    presence of, 99
    putting him first in our lives, 131
    reaction to global depravity, 128
    as savior, 79
    and scientific method, 16–17
    timelessness of, 77–78
    and unconditional love, 18–19, 37
    voice of, 99, 142–43, 144, 145, 146, 148, 149–53
good deeds, 62
gravity, 15, 55–60, 62, 65, 68, 86, 93–94, 138–39
    law of, 15, 90, 93
Great Leap Forward, 162–63, 169
Great Year, 31
Greece, classical, 31
"Greek Miracle," 13–14
Groundhog Day (movie), 25

hard sciences, 103, 105
heaven, 37, 61
hedonism, 32
heresy, 32
hidden reality
    and the Bible, 60–64
    practical application in everyday life, 65–69
    and science, 53–60

SUBJECT INDEX

Hinduism, 18, 27–28, 37, 46, 62
His Master's Voice (painting), 144–45
holiness, 170
holy trinity, 93
Homo sapiens, 156
Homo sapiens sapiens, 163, 166, 168
hope, 134
humans, uniqueness of, 154–72
    and the Bible, 163–69
    and science, 156–63
    what it means in everyday life, 169–72
humility, 45, 49, 50, 97, 128

image of God, 37, 166, 167
incompleteness theorem, 105, 110
India, 27
inertia, 75
inflation theory, 15, 89
innovation, 167
instantaneity, 140–41
instantaneous communication, 135–53
    and the Bible, 142–48
    and science, 136–42
    what it means in everyday life, 148–53
Instinct (Jakes), 18
intelligence, 161, 167
intelligent design, 69
interactions, 137–38, 140–41, 142, 146, 147
intuition, 102, 106, 109–13
invertebrates, 165
invisible realm, 163
Islam, 111

Jehovah's Witnesses, 46

Jesus
    asked about truth, 11
    birth of, 79–80, 143
    contradictory nature of, 44–49
    emotionality of, 48–49
    life of, 29
    and light, 76–80
    resurrection of, 63
Jews/Judaism, 18, 28, 45, 46, 50, 62, 111, 170
John Templeton Foundation, 145–46

karma, 26, 28, 37
kingdoms, 155
kosmos. *See* cosmos

Last Supper, 20–21
Last Temptation of Christ, The (movie), 45
laws, 15
life after death. *See* afterlife
Life After Death (Chopra), 63
light, 54, 57, 70–83, 163. *See also* darkness
    barrier, 74
    and the Bible, 76–80
    quanta, 43
    and Satan, 79
    and science, 71–76
    speed of, 15, 71–75, 140–41
    and time, 76–77
    in vacuum, 15, 72, 73
    waves, 43, 54
    what it means in everyday life, 80–83
linear time, 24–37
    and the Bible, 27–30
    and objective truth, 34
    and science, 30–34

what it means in everyday life, 34–37
local interactions, 137–38, 142
logic, 23, 67, 82, 103–5, 106–8, 109, 110–13, 114, 116, 117, 122, 148–49, 167
Lucifer Effect, The (Zimbardo), 170

magnetism, 138
Manhattan Project, 75
Manichaeism, 78, 111
marginal children, 114
mass, 58–59, 75–76
material reality, 63
mathematics, 102–3, 108–9
meaning, absolute, 12
mechanics, Newtonian, 120
messianic Jews, 45
metaphysics, 18
meteorites, 70–71
Meteorologica (Aristotle), 32
method
    religious, 111, 112, 113, 172
    scientific. *See* scientific method
Milky Way, 56
Million Monkey project, 98
Minds of Billy Milligan, The (Keyes), 39
miracles, 150–52
missing mass problem, 58
mitzvoth, 62
More Beautiful You (song), 37
Muslims, 46

natural law, 98
natural selection, 93, 94, 156, 167
Nazis, 170
nephesh, 165
New Testament, 28–30, 117, 129, 163

Nicene Creed, 47
1984 (Orwell), 133
nonlocal interactions, 137, 138, 139, 141, 142, 146, 147
nothingness, 54, 89–90, 97, 99

objective truth, 11–23, 105
    belief in its existence, 23
    and the Bible, 17–21
    and Christianity, 20
    and linear time, 34
    respect for, 22
    and science, 13–17
    what it means in everyday life, 21–23
    why it matters, 12
Old Testament, 29, 129, 163
omnipresence, 146–47
"One Solitary Life" (Francis), 129–30
opposites, 40, 106. *See also* contradictory nature
origin
    godly, 17
    of Israel, 29
    of light, 71
    of religions, 27, 78
    theory, 88, 89
    of times, 30, 33
    of universe, 85–91, 93, 106
Our Family Tree (Radko), 132–33
oxygen, 54

pair annihilation, 76, 80
pair creation, 76
paradox, 104
particles
    light, 140
    subatomic, 74
    and waves, 40–44

SUBJECT INDEX

persistence of consciousness, 63
Philanthropy Project, 145–46
phlogiston, 53–54
photoelectricity, 42–43
photons, 140, 141
physical constants, 15
physical-material phenomena, 16
physics, 15, 41, 102–3, 120, 121
plane geometry, 14, 104
plants, 163, 164
polarization, 140
positrons, 76
"Power of Words" (Poe), 124
primeval atom, 88
producers, 163, 164
proof, 103–5. *See also* certainty and uncertainty
Proslogion (Anselm), 109

quantum entanglement, 139–41
quantum mechanics, 137, 139–40, 147
quantum theory, 41–42, 47, 49–50
quantum vacuum, 60, 89–90, 93, 99, 168
quintessence, 60

randomness, 93–94, 95, 98, 125, 131
rational-materialism, 17–18
reality
    extremities of, 49
    hidden (*See* hidden reality)
    material vs. spiritual, 63
    physical, 127
reason, 161
reasoning, deductive, 103, 107–8
reincarnation, 27–28, 31
relativity, theories of, 14–15, 33, 55–60, 72–73, 74, 75, 86, 138–39, 140

resonance, 152
resurrection, bodily, 63, 64
revelation, divine, 18, 110, 112

SAD (seasonal affective disorder), 81
samsara, 27–28
Satan, 50–51, 78, 79, 151
science
    cause and effect, 120–26
    and certainty and uncertainty, 102–8, 109
    and contradictory nature, 40–44
    and hidden reality, 53–60
    and humans' uniqueness, 156–63
    and instantaneous communication, 136–42
    and light, 71–76
    and linear time, 30–34
    and objective truth, 13–17
    origin of the universe, 85–91
scientific method, 16–17, 34, 68, 105, 111, 112, 113, 154, 172
"seeing is believing," 52, 67, 69
self-deception, 150
selflessness, 167
sensory perception, 110
silence, 98
simplicity, 125
sin, 48
soft sciences, 103, 105, 106
solar system, 56
Solvay International Conference, 87
Son of God, 81
soul, 28, 61, 63, 164–66. *See also* afterlife
soulish beings, 165, 166, 168
soulless beings, 164
sound, 136–37

sound barrier, 74
"Sound of Thunder, A" (Bradbury), 124–25
space-time topography, 138–39
speed of light, 15, 71–75, 140
spirituality, 163, 164–65, 168, 169
spiritual quotient (SQ), 167, 168
spiritual reality, 63
starlight, 54, 57
steady state hypothesis, 85–86, 88–89, 91
Stoicism, 31–32
strange attractor, 126
Summa Theologica (Aquinas), 112
sunshine, 80–81. *See also* light
supernovae, 59
syllogism, 103
sympathetic vibration, 152

tachyons, 74
tardyons, 73, 74–75, 76
theodicy, 82
theology, 108, 112
theories, 15–16, 105
theory of everything, 139
They All Laughed (Flatow), 109
Through the Looking Glass (Carroll), 49
Tibet, 26
time
    circular, 25–26, 27–28, 30, 31–32, 35
    and light, 74–75, 76–77
    linear (*See* linear time)
Time in History (Whitrow), 27
Time Machine, The (Wells), 34–35
topology, 113–14
Torah, 164
transfiguration, 143

transmigration (of soul), 28
Trinity, 93
trust, 112–13
truth, 106, 113, 149, 169, 171. *See also* objective truth
Two New Sciences (Galileo), 71

uncertainty. *See* certainty and uncertainty
uncertainty principle, 105–6, 110
unconscious will, 65
uniqueness, humans'. *See* humans, uniqueness of
universe, 14. *See also* cosmos
    acceleration of, 67
    Bible version of origin of, 91–96
    expansion of, 57–58, 59, 86–90
    linear, 121–22, 123
    origin of, 84–99, 106
    physical-material behavior of, 16
    science of the origin of, 85–91
    visible, 60
    what theories of origin of mean, 97–99

Vedism, 27
vertebrates, 165
voice: *See* God: voice of

waves, 40–44, 54
weather, 122–23
What Jesus Means to Me (Gandhi), 46
will, 65
wisdom, 33, 47, 49, 50, 67, 109, 117, 163

Zoroastrianism, 78

ABOUT THE AUTHOR

A three-time Emmy Award winner, best-selling author, and former Harvard University instructor, Dr. Michael Guillen, PhD, is known and loved by millions as the ABC News science editor, a post he filled for fourteen years (1988–2002). In that capacity, he appeared regularly on *Good Morning America*, *20/20*, *Nightline*, and *World News Tonight*. He is host of *Where Did It Come From?*, a popular one-hour primetime series for the History Channel that debuted in 2006.

In association with Anonymous Content, Dr. Guillen produced *Little Red Wagon*, the award-winning motion picture written by Patrick Sheane Duncan (*Mr. Holland's Opus*) and directed by David Anspaugh (*Rudy*, *Hoosiers*). He is also a columnist for *U.S. News and World Report* and *Fox News*.

Over the years, Dr. Guillen's articles have appeared in distinguished publications such as *Science News*, *Psychology Today*, the *New York Times*, and the *Washington Post*. He is the author of the bestselling *Bridges to Infinity: The Human Side of Mathematics* and *Five Equations That Changed the World: The Power and Poetry of Mathematics*. His book *Can a Smart Person Believe in God?* tells of his lifelong attempt to reconcile science with religion.

A native of East Los Angeles, Dr. Guillen earned his BS from UCLA and his PhD in physics, mathematics, and astronomy from Cornell University. He has received honorary doctorates from the University of Maryland and Pepperdine University. In 2000 he was elected to the renowned, century-old Explorers Club.

Dr. Guillen is a popular speaker and serves as president of Spectacular Science Productions, Inc., and Filmanthropy Media Incorporated. He also is chairman and president of Philanthropy Project. For more, go to www.michaelguillen.com.